THE LITTLE ENGINEER COLORING BOOK

SPACE & ROCKETS

SETH MCKAY

The Little Engineer Coloring Book: Space & Rockets by Seth McKay
www.TheLittleEngineerBooks.com

Copyright © 2019 by Seth McKay

All rights reserved. No portion of this book may be reproduced, stored in a retrieval system, or transmitted in any form or by any meanselectronic, mechanical, photocopy, recording, scanning, or other-except for brief quotations in critical reviews or articles, without the prior written permission of the publisher.

Creative Ideas Publishing titles may be purchased in bulk for educational, business, fund-raising, or sales promotional use. For information, please email permissions@TheLittleEngineerBooks.com.

ISBN-13: 978-1-952016-04-2

Published by: Creative Ideas Publishing

Table of Contents

Introduction (*for Parents*) v
Introduction (*for the Little Engineers*) vi
We Are Explorers 1
Earth 2
The Solar System 3
The Sun 4
The Moon 5
How Do We Learn About Space? 6
Telescope 7
Observatory 8
Satellites 9
Space Probes 10
Rovers 11
Astronauts 12
Rockets 13
Fly Around the Earth 14
Going to the Moon 15
Saturn V 16
Command Module 17
Launch Escape System 18
Launch Stages 19
Going to the Moon: Step 1 20
Going to the Moon: Step 2 21
Going to the Moon: Step 3 22
Going to the Moon: Step 4 23
Going to the Moon: Step 5 24
Going to the Moon: Step 6 25

Going to the Moon: Step 7	26
Going to the Moon: Step 8	27
Going to the Moon: Step 9	28
Going to the Moon: Step 10	29
Going to the Moon: Step 11	30
Going to the Moon: Step 12	31
Going to the Moon: Step 13	32
Going to the Moon: Step 14	33
Going to the Moon: Step 15	34
Going to the Moon: Step 16	35
Going to the Moon: Step 17	36
Going to the Moon: Step 18	37
Going to the Moon: Step 19	38
Space Shuttle	39
Space Shuttle: Side Boosters	40
Space Shuttle: Cargo Bay	41
Space Shuttle: Return Home	42
SpaceX: Re-Usable Rockets	43
SpaceX: How the Rocket Lands	44
Modern Rockets	45
Next Big Adventure	46
Revisit the Moon	47
Journey to Mars	48
Launch More Space Probes	49
Why Explore Space	50
Certificate of Completion	51
BONUS: Special Preview of Cars and Trucks Coloring Book	52

Introduction *(for parents)*

Thank you so much for your interest in this book. In this series of books, your child will be exposed to new, interesting topics and begin to understand how things work.

I've always wanted my children to understand how things work, and in general, understand that objects are not magical boxes that work for some unknown reason. I want them to understand that objects are quite simple when broken down into a few key components. Without any direction, it is possible for a child to get the wrong understanding of objects that we consider to be simple.

Personally, I find space incredibly interesting, and I think we are on the cusp of some great leaps in space exploration. SpaceX and Blue Origin are both releasing super heavy lift rockets that are re-usable and will plummet the cost of space travel. We will surely see some very exciting things happen in space in the next 20 years.

The young ones coloring this book will hopefully get to experience a lifetime of leaps and bounds in space exploration.

I chose to make this book a coloring book because I wanted to ensure that the book was also fun! I don't want this book to be too serious, boring or overwhelming. Children enjoy coloring so this book helps incorporate some new topics to learn while also not being too serious.

Book Objective:

The objective is simply to begin to expose your child to new, interesting concepts that might catch his or her attention and help cultivate curiosity in how things work.

Introduction *(for the Little Engineers)*

Hello Little Engineer!

My name is Seth, and I'm the Chief Engineer of this book. I'm so happy you have this book and we will be sure to have loads of fun.

Our solar system is such a cool, interesting place and even grown ups are learning new things about space every day!

In fact, we are going to need your help making new discoveries because there are still so many things we don't know about space.

Enjoy coloring, ask your parent or teacher lots of questions, and most of all, have fun!

WE ARE EXPLORERS!

WE, HUMANS, LIKE TO EXPLORE! HUNDREDS OF YEARS AGO, EXPLORERS WOULD GET ON BOATS AND SPEND MONTHS ON A BOAT AS THEY SAILED ACROSS OCEANS TO EXPLORE THE WORLD. SOON, WE MAY DO THE SAME THING IN SPACE AS WE EXPLORE THE SOLAR SYSTEM.

EARTH

THIS IS OUR HOME. WE ARE ALL ON EARTH TOGETHER, AND IT IS A PERFECT PLANET FOR US TO LIVE.

THE SOLAR SYSTEM

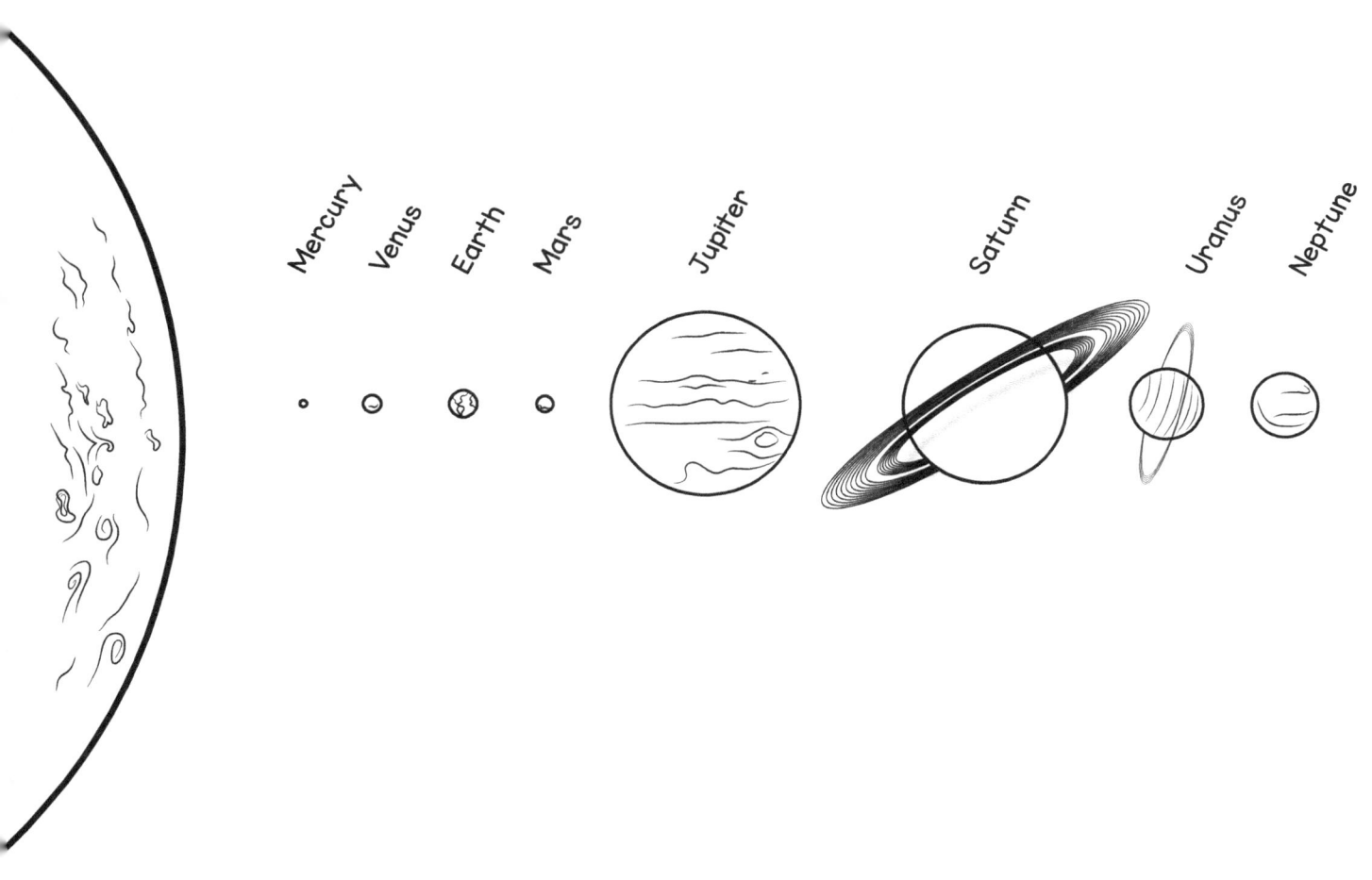

THE EARTH IS IN THE SOLAR SYSTEM. OUR SOLAR SYSTEM CONSISTS OF THE SUN AND 8 PLANETS.

THE SUN

THE SUN IS AT THE CENTER OF OUR SOLAR SYSTEM. ALL THE PLANETS ORBIT AROUND THE SUN. THE EARTH IS VERY SMALL COMPARED TO THE SUN.

THE MOON

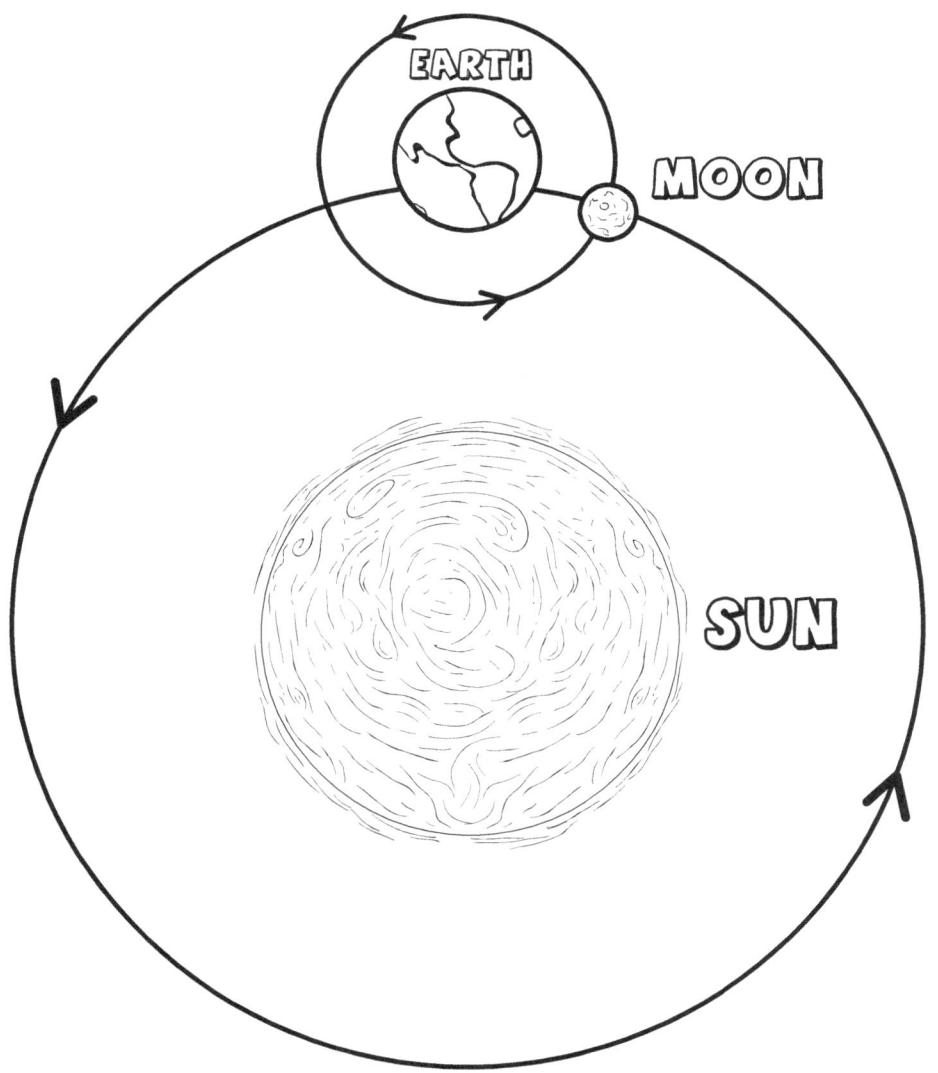

**THE EARTH ORBITS AROUND THE SUN,
AND THE MOON ORBITS AROUND THE EARTH.**

HOW DO WE LEARN ABOUT SPACE?

SPACE IS SO AMAZING AND INTERESTING! TELESCOPES, SATELLITES, ROVERS, PROBES AND HUMAN SPACE TRAVEL ARE THINGS THAT HELP US LEARN ABOUT SPACE.

TELESCOPES

TELESCOPES HAVE MADE MANY SPACE DISCOVERIES FOR US. EVEN A SMALL TELESCOPE CAN SEE JUPITER FROM EARTH. THE FIRST TELESCOPE WAS MADE 400 YEARS AGO.

OBSERVATORY

AN OBSERVATORY IS A GIANT TELESCOPE THAT IS BUILT INTO A BUILDING.

SATELLITES

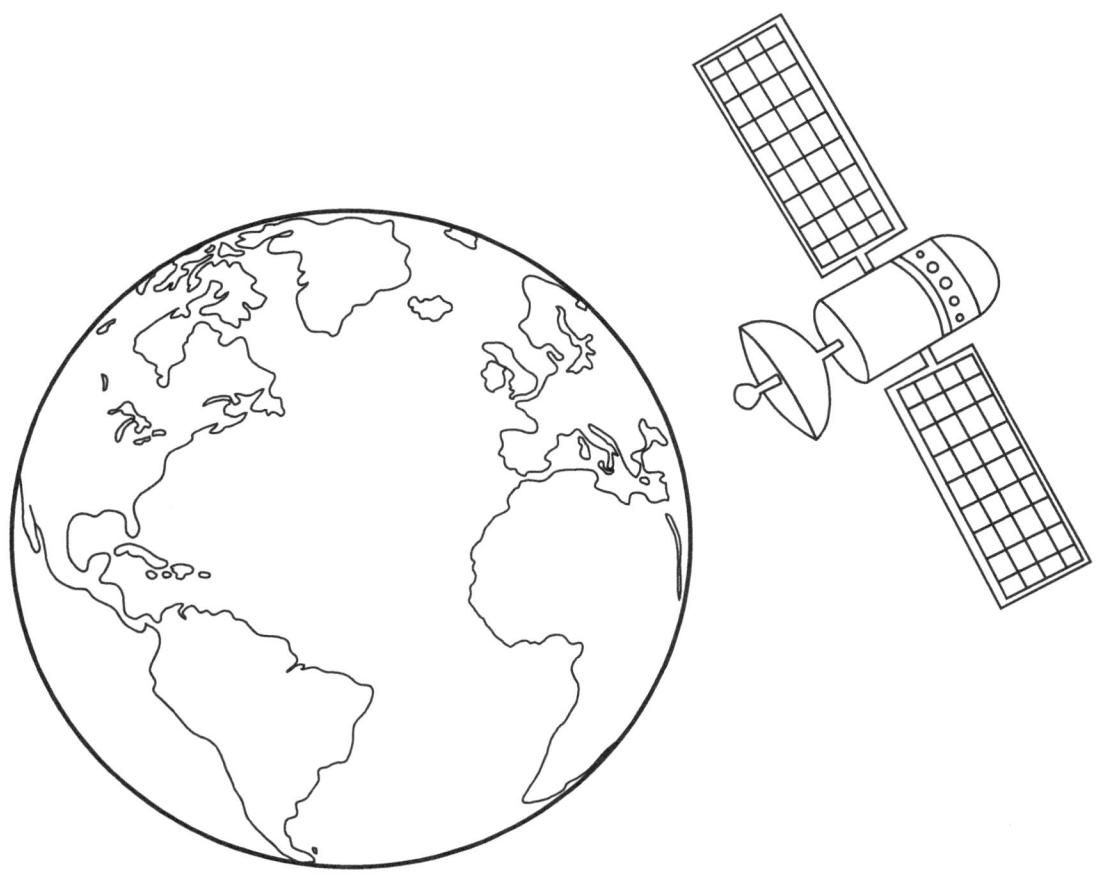

A SATELLITE IS AN OBJECT IN ORBIT AROUND A PLANET. MANY SATELLITES ORBIT EARTH. SATELLITES ARE ALSO IN ORBIT AROUND OTHER PLANETS TO HELP US LEARN ABOUT THE PLANETS.

SPACE PROBES

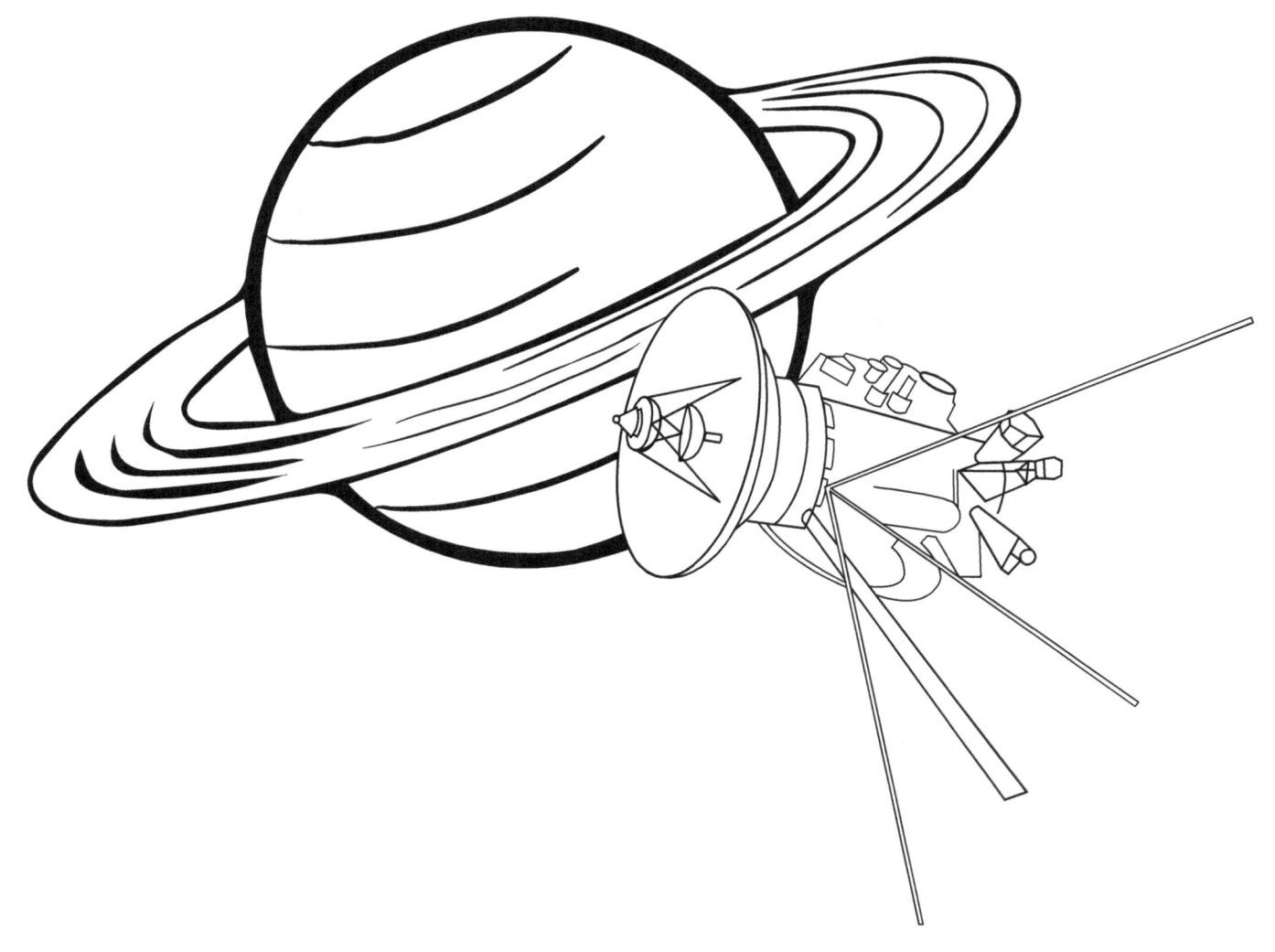

A SPACE PROBE IS SIMILAR TO A SATELLITE EXCEPT PROBES MAY NOT STAY IN ORBIT. INSTEAD, THEY MAY SIMPLY FLY BY A PLANET TO LEARN ABOUT IT OR FLY DEEP INTO SPACE.

ROVERS

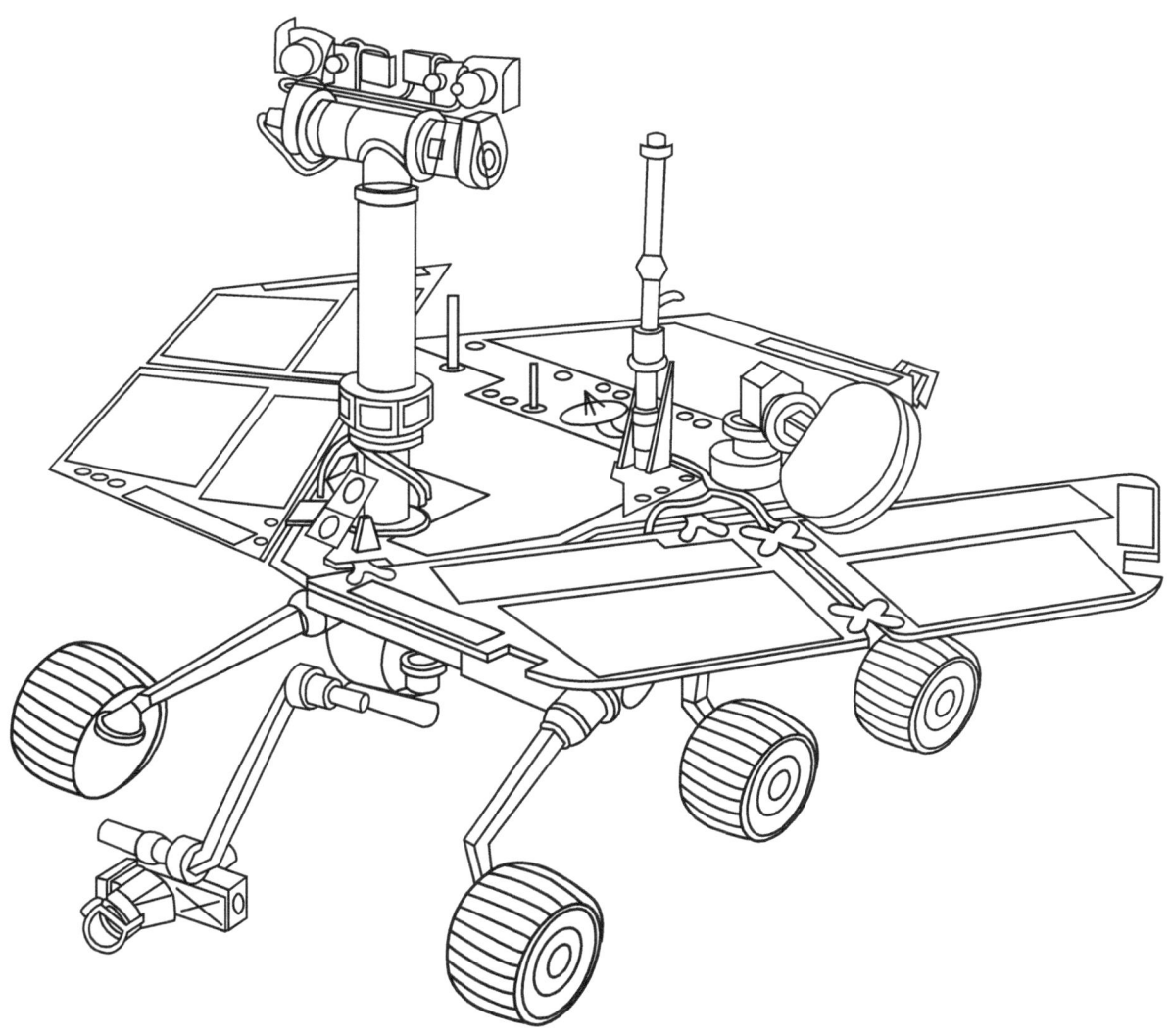

ROVERS ARE SCIENCE INSTRUMENTS LIKE A SPACE PROBE EXCEPT ROVERS WILL LAND ON THE PLANET AND EXPLORE THE PLANET AS IT DRIVES AROUND.

ASTRONAUTS

A SPACE EXPLORER IS CALLED AN ASTRONAUT. THEY ARE TRAINED TO LIVE IN SPACE. THEIR SUITS MAKE SURE THEY HAVE AIR TO BREATHE BECAUSE SPACE DOESN'T HAVE OXYGEN LIKE ON EARTH.

ROCKETS

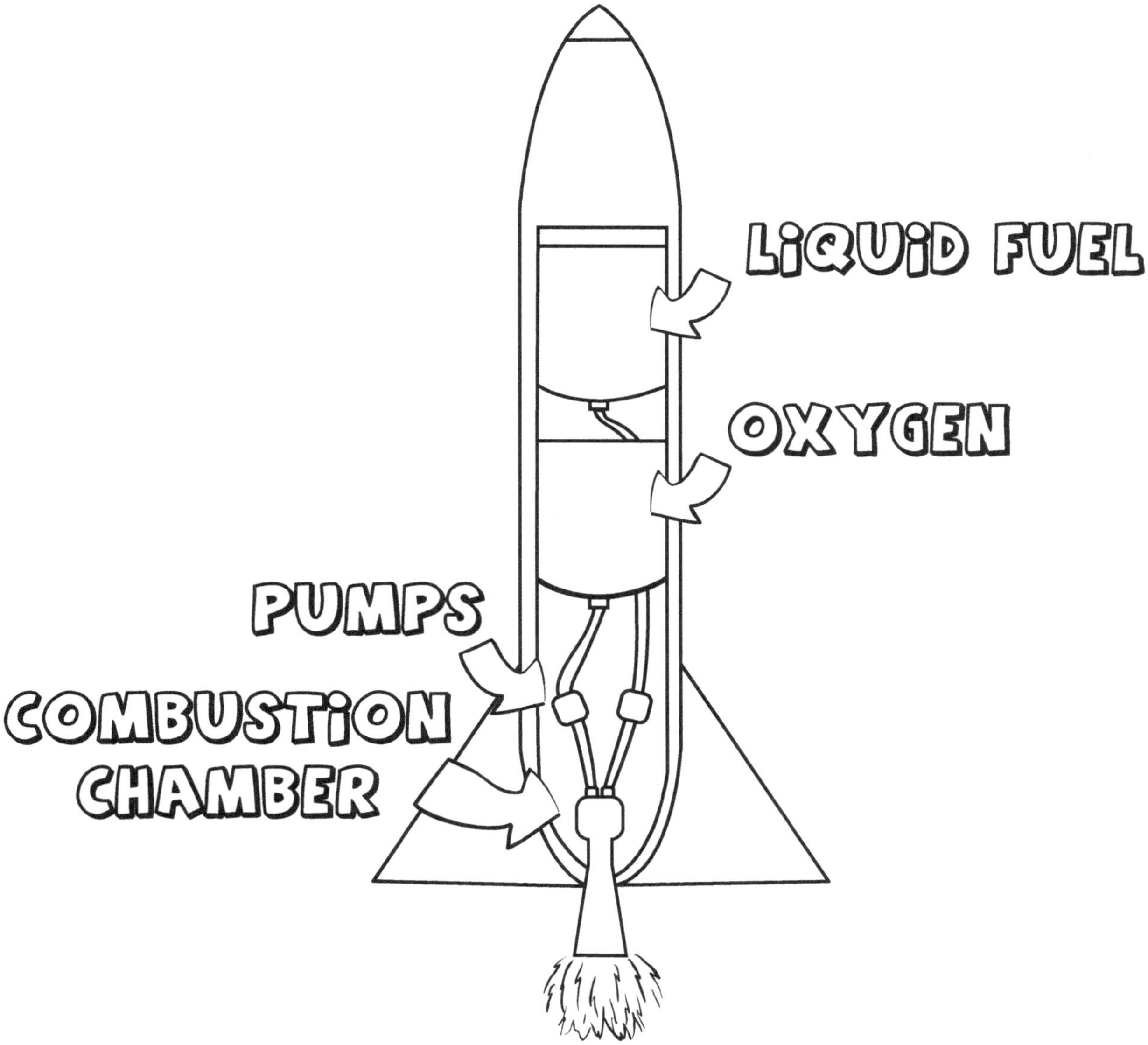

Labels on diagram: LIQUID FUEL, OXYGEN, PUMPS, COMBUSTION CHAMBER

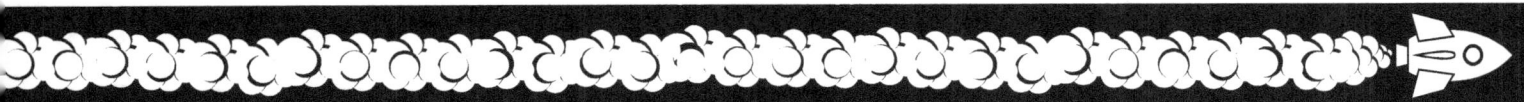

GIANT ROCKETS, BIGGER THAN YOUR HOUSE, ARE USED TO LAUNCH ASTRONAUTS OR SATELLITES INTO SPACE. ROCKETS BURN FUEL AND BLAST THE EXHAUST GAS DOWN SO THE ROCKET WILL GO UP.

The Little Engineer Coloring Book: Space & Rockets

FLY AROUND THE EARTH

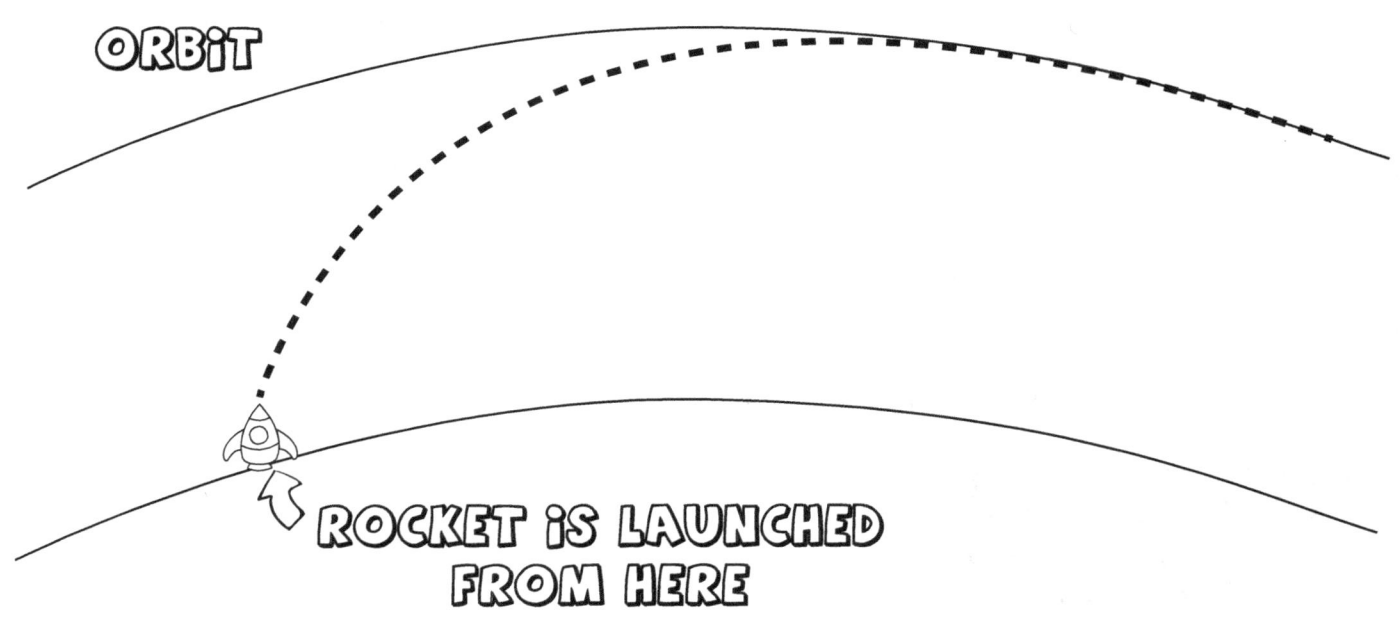

ROCKETS DON'T FLY STRAIGHT UP TO GET INTO SPACE. THEY FLY UP AND THEN SIDEWAYS TO GET INTO ORBIT AROUND EARTH.

GOING TO THE MOON

HUMANS TRAVELED TO THE MOON FOR THE FIRST TIME 50 YEARS AGO. A NEW ROCKET AND MANY NEW TECHNOLOGIES WERE DEVELOPED TO GET TO THE MOON.

SATURN V

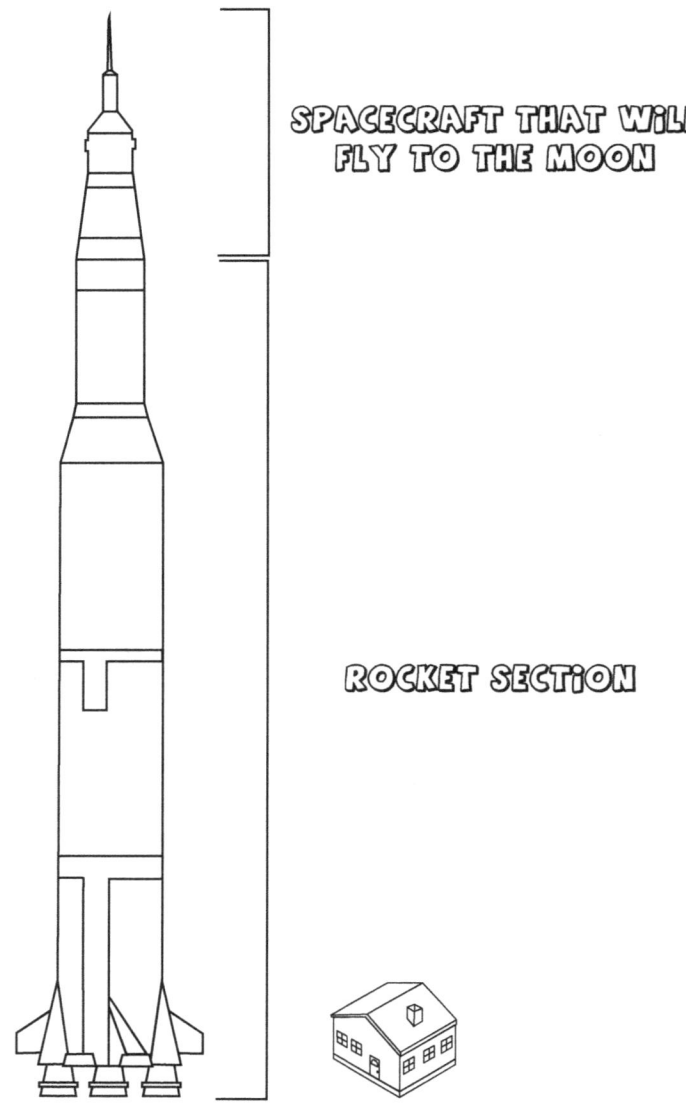

SPACECRAFT THAT WILL FLY TO THE MOON

ROCKET SECTION

THE SATURN V (FIVE) IS AN INCREDIBLE ROCKET. IT WAS THE ROCKET THAT TOOK MEN TO THE MOON. LOOK HOW BIG IT IS COMPARED TO A HOUSE!

COMMAND MODULE

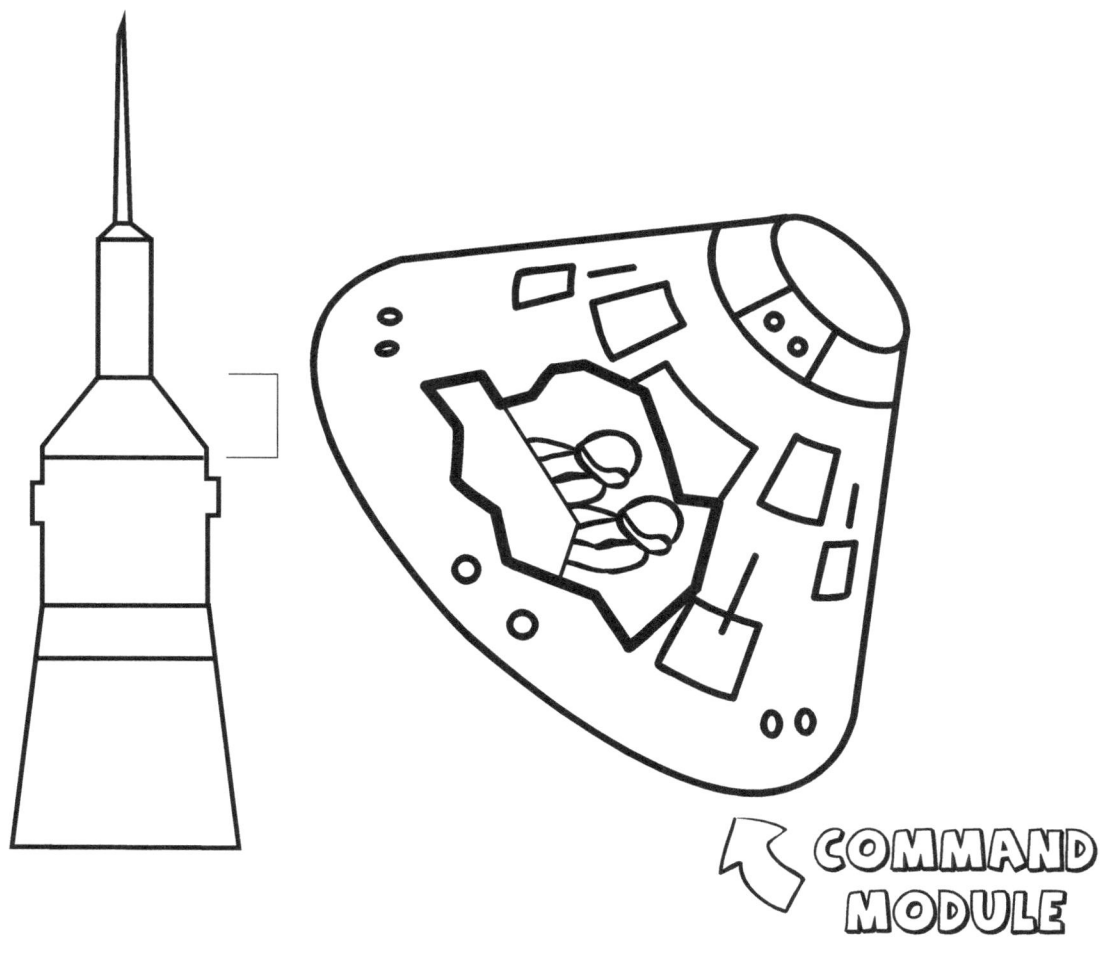

THE SMALL COMMAND MODULE ON TOP OF THE ROCKET IS THE ONLY PART OF THE HUGE ROCKET THAT RETURNS TO EARTH.

LAUNCH ESCAPE SYSTEM

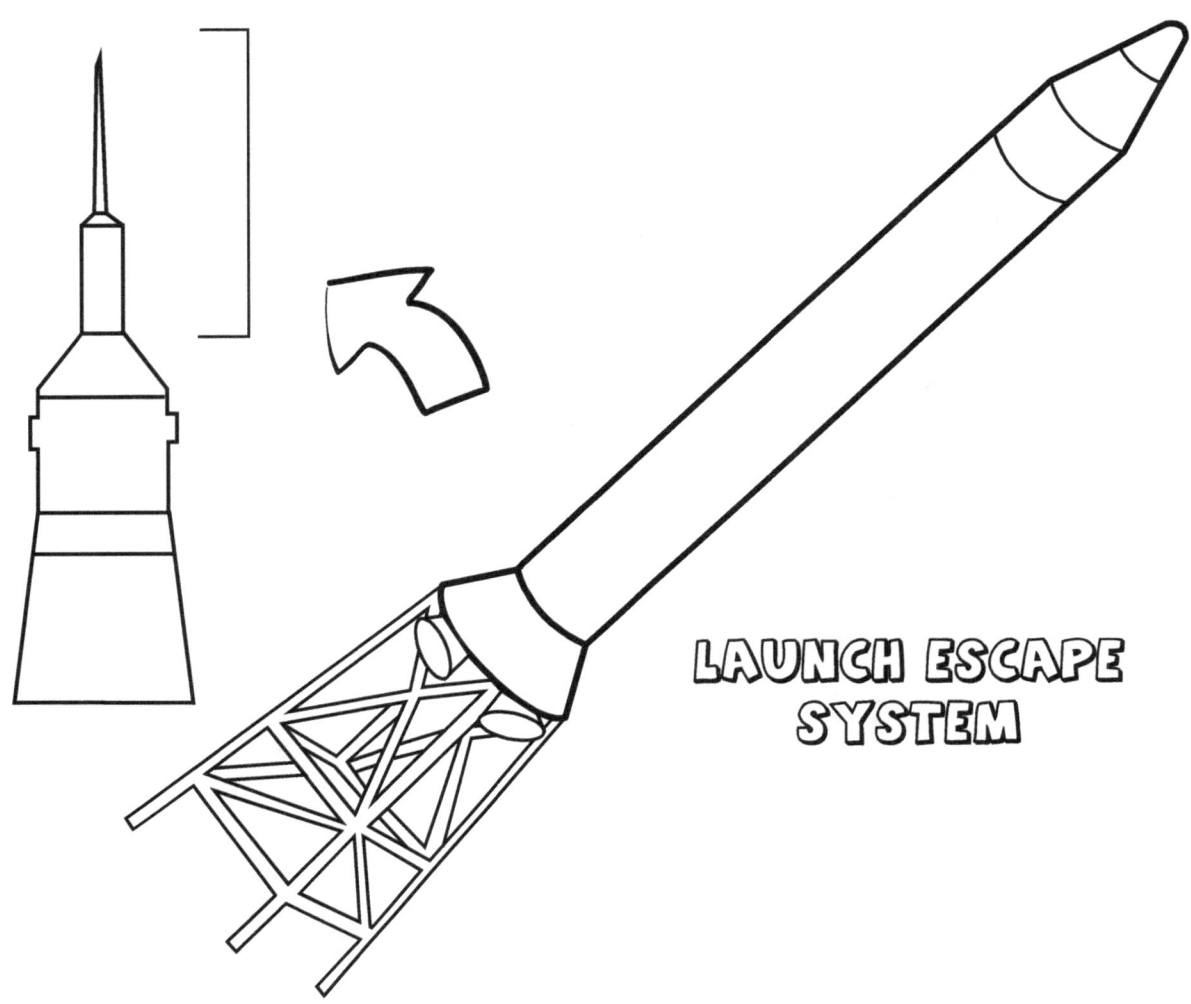

LAUNCH ESCAPE SYSTEM

THE TOP OF THE ROCKET HAS SOME EXTRA SMALL ROCKETS THAT WILL PULL THE COMMAND MODULE AND THE ASTRONAUTS AWAY TO SAFETY IF THE MAIN ROCKET HAS A PROBLEM.

LAUNCH STAGES

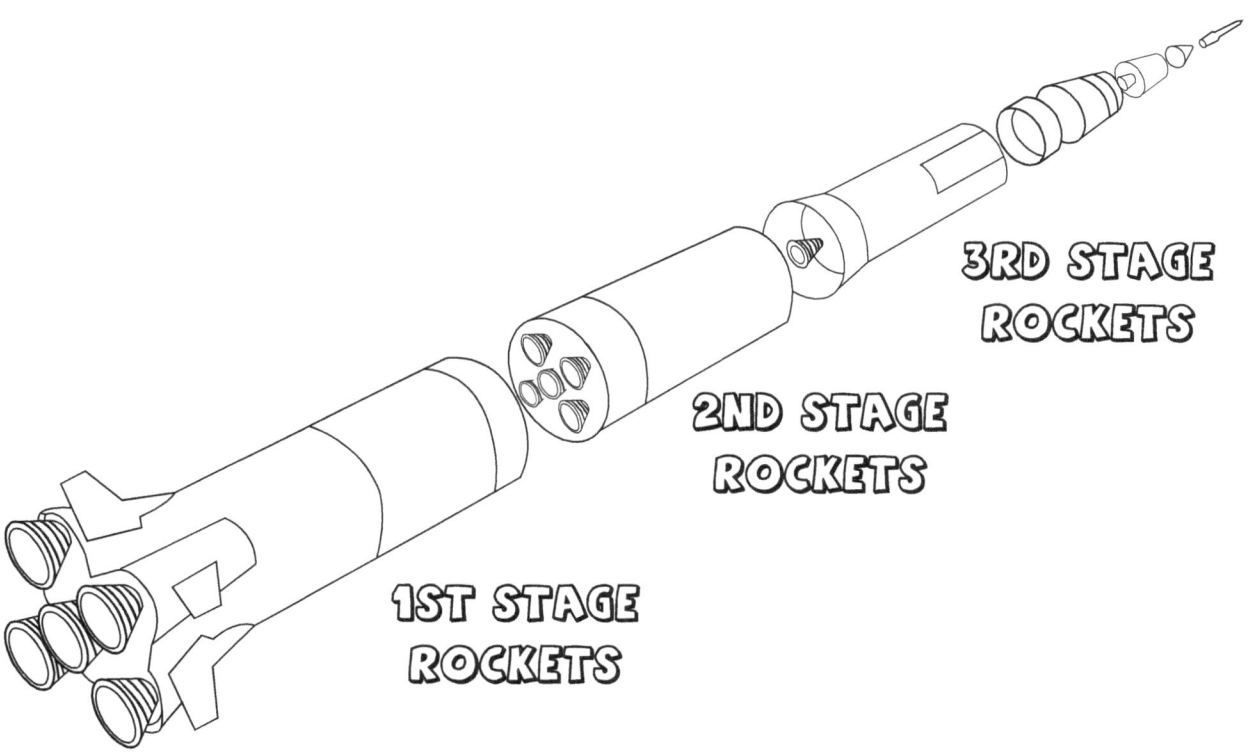

THE ROCKET HAS SEVERAL DIFFERENT SECTIONS CALLED STAGES.

GOING TO THE MOON STEP 1

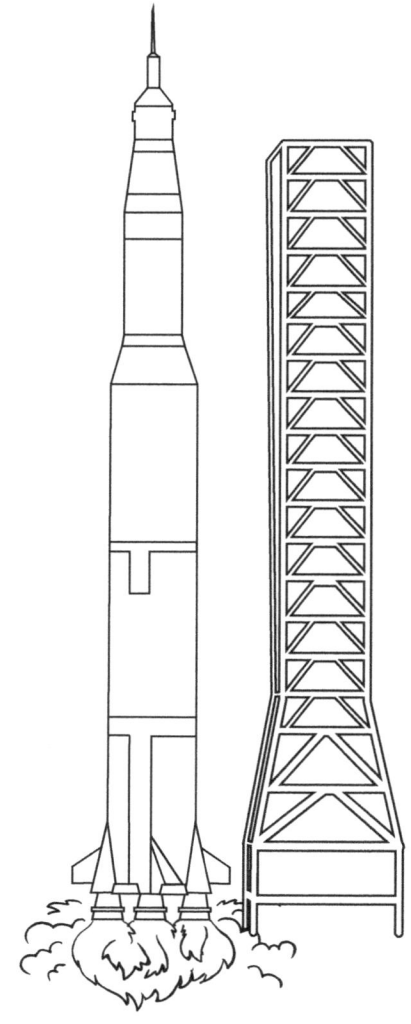

STEP 1: LAUNCH THE GREAT SATURN V FROM THE LAUNCH PAD.

GOING TO THE MOON
STEP 2

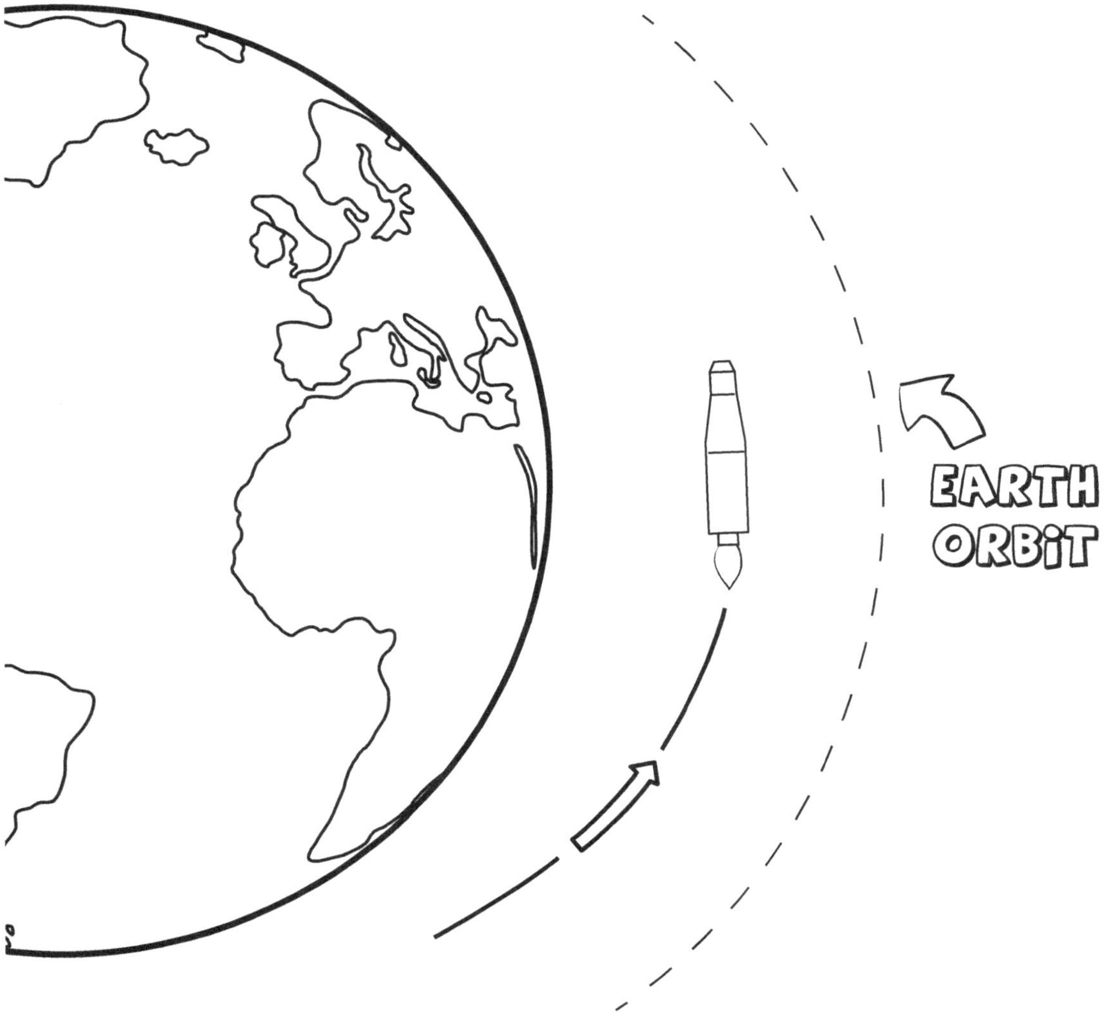

EARTH ORBIT

STEP 2: STAGE 1 AND STAGE 2 OF THE ROCKET DROP AWAY DURING LAUNCH AS THE ROCKET GOES INTO ORBIT AROUND THE EARTH.

The Little Engineer Coloring Book: Space & Rockets

GOING TO THE MOON
STEP 3

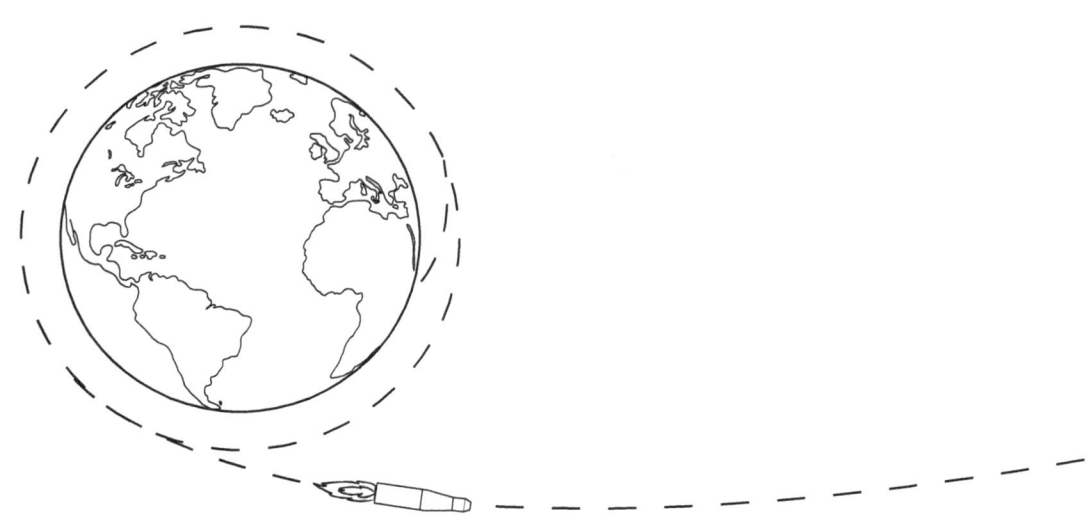

STEP 3: FIRE STAGE 3 OF THE ROCKET TO LEAVE EARTH'S ORBIT AND HEAD TOWARD THE MOON.

GOING TO THE MOON STEP 4

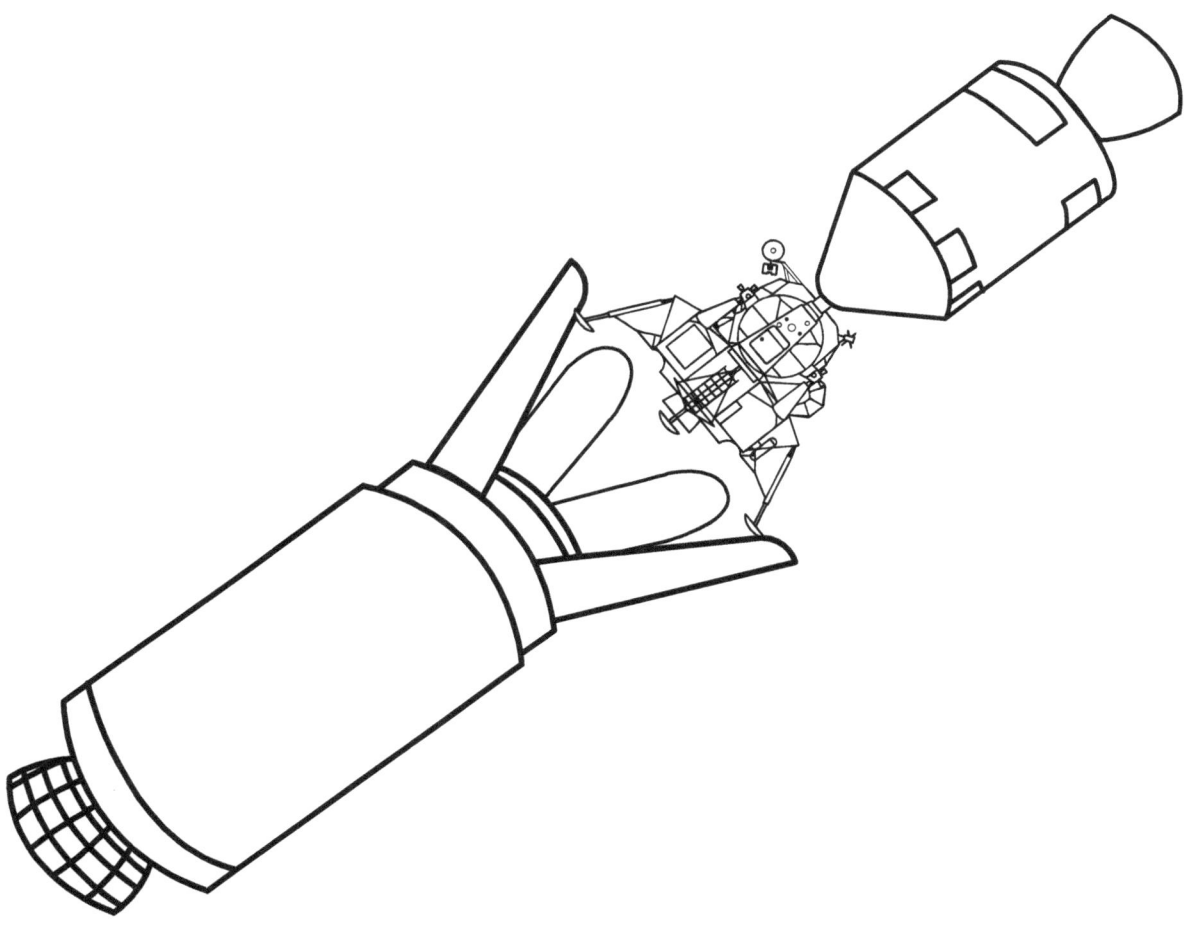

STEP 4: ONCE THE SPACECRAFT REACHES THE CORRECT SPEED, STAGE 3 OF THE ROCKET SEPARATES FROM THE LUNAR MODULE.

GOING TO THE MOON
STEP 5

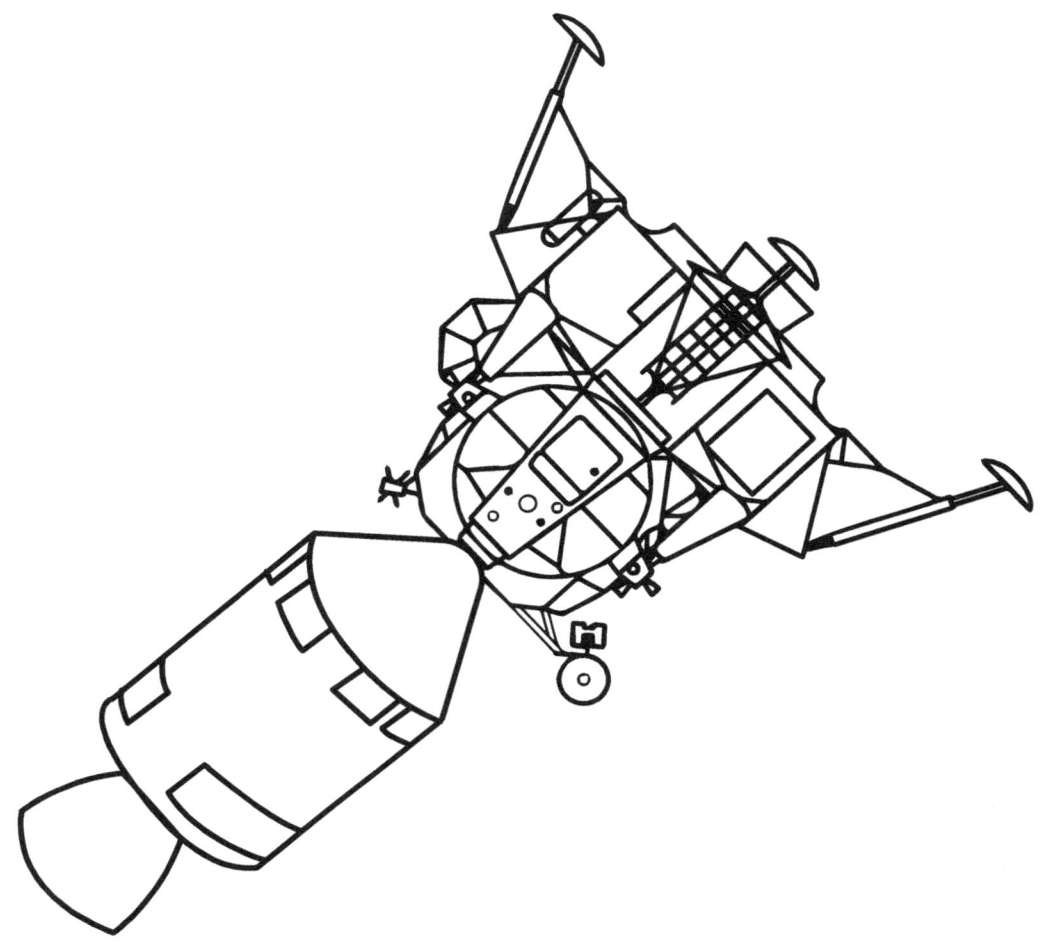

STEP 5: THE LUNAR MODULE IS NOW ALL THAT IS LEFT OF THE SPACECRAFT, AND IT CONTINUES TOWARDS THE MOON.

GOING TO THE MOON
STEP 6

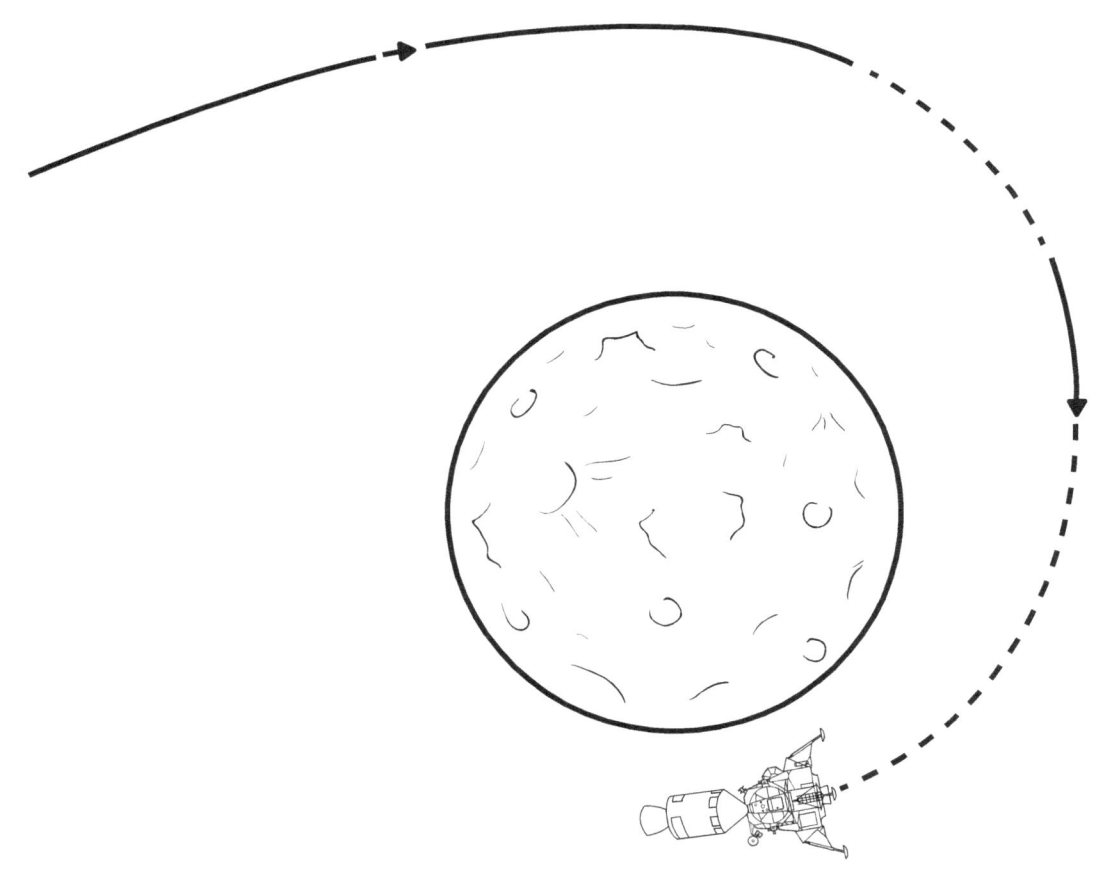

STEP 6: ONCE IT'S NEAR THE MOON, THE LUNAR MODULE USES ITS ROCKET TO BLAST BACKWARDS AND SLOW DOWN TO STAY IN AN ORBIT AROUND THE MOON.

GOING TO THE MOON
STEP 7

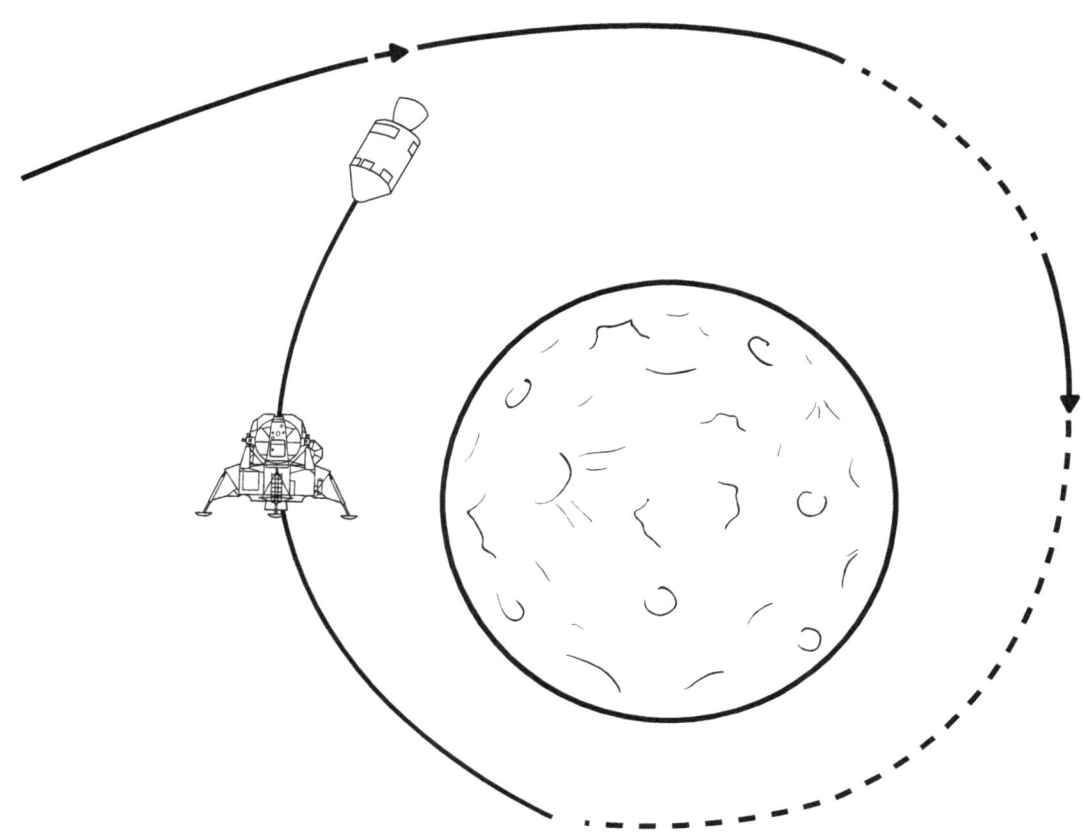

STEP 7: THE LUNAR LANDER THEN SEPARATES FROM THE COMMAND SERVICE MODULE.

GOING TO THE MOON STEP 8

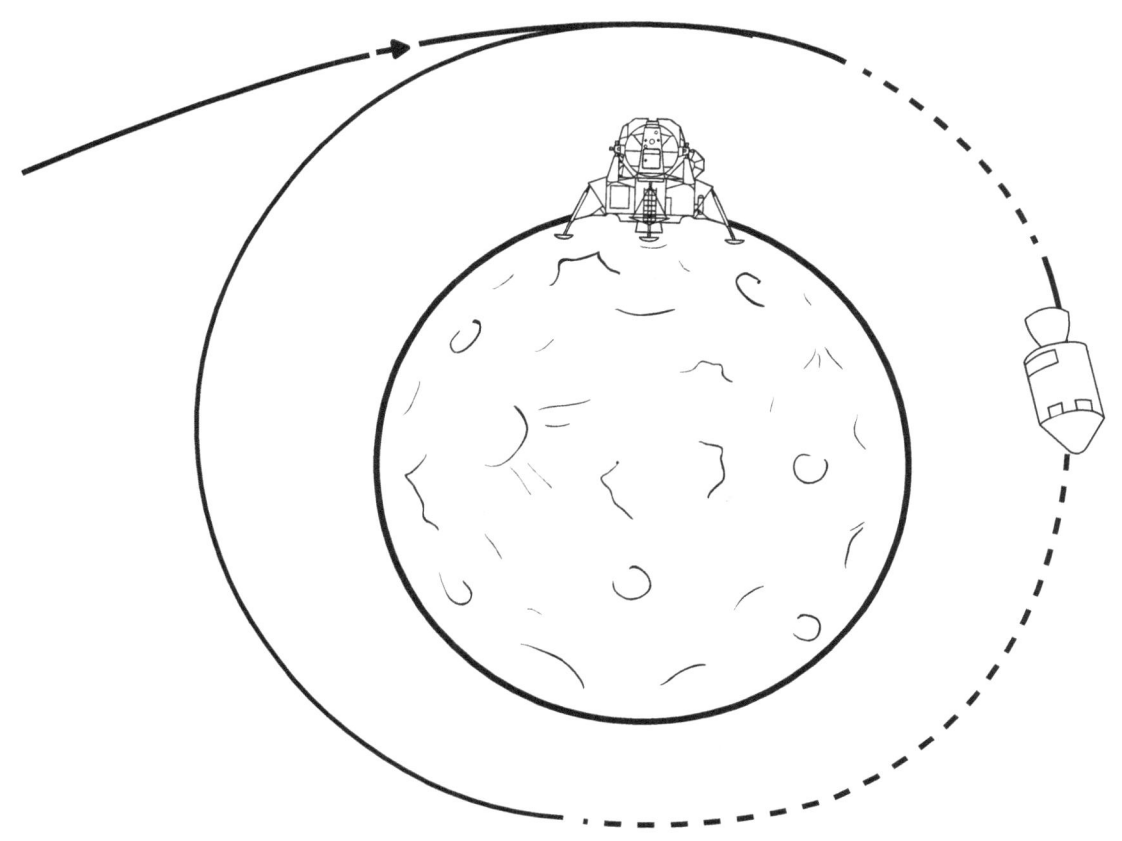

STEP 8: WHILE THE COMMAND SERVICE MODULE CIRCLES THE MOON, THE LUNAR LANDER USES ITS ROCKETS TO SLOW DOWN AND LAND SOFTLY ON THE MOON.

GOING TO THE MOON
STEP 9

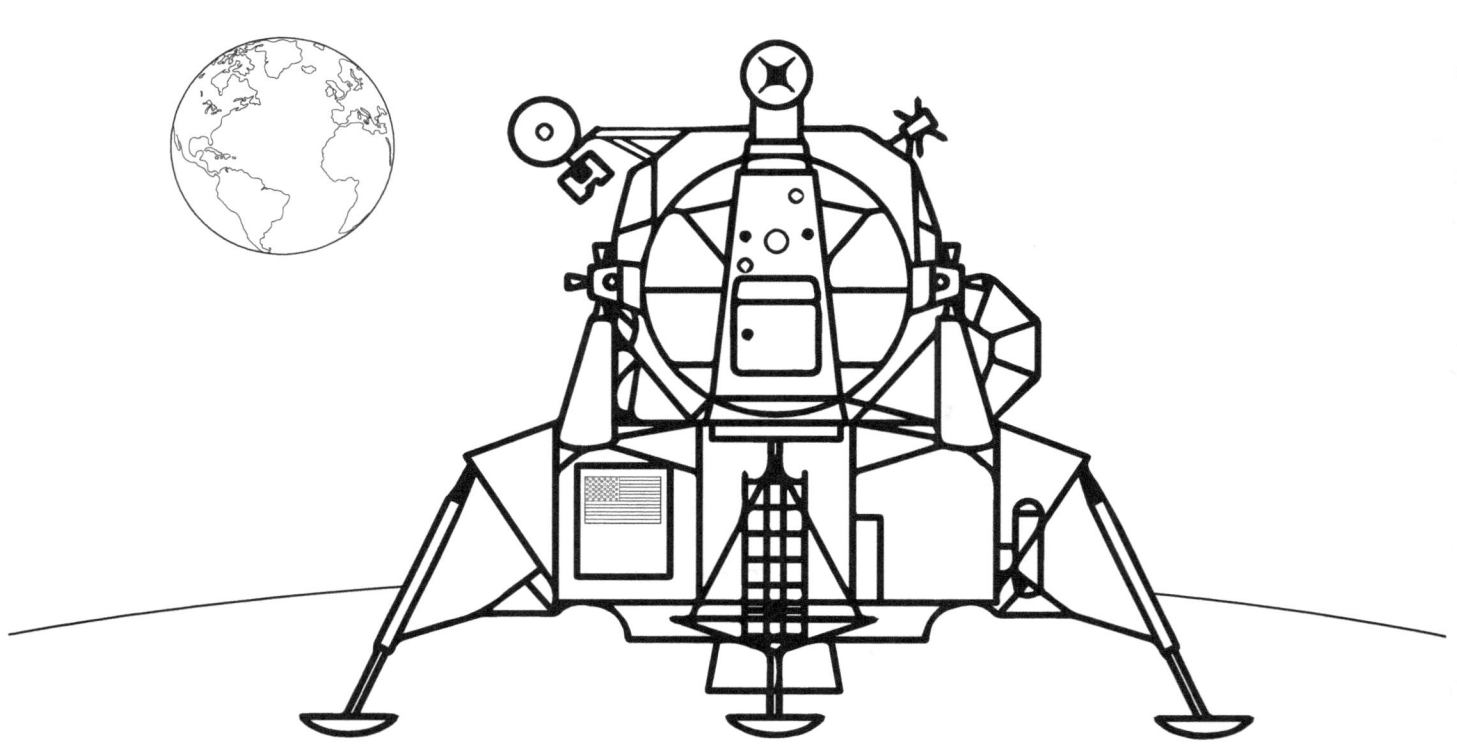

STEP 9: THE LUNAR LANDER IS ON THE MOON! ASTRONAUTS ARE NOW 240,000 MILES AWAY FROM EARTH AS THEY EXPLORE THE MOON!

The Little Engineer Coloring Book: Space & Rockets

GOING TO THE MOON STEP 10

STEP 10: ASTRONAUTS CONDUCT EXPERIMENTS AND COLLECT ROCK SAMPLES FROM THE MOON SO WE CAN LEARN MORE ABOUT THE MOON AND HOW IT WAS FORMED.

GOING TO THE MOON
STEP 11

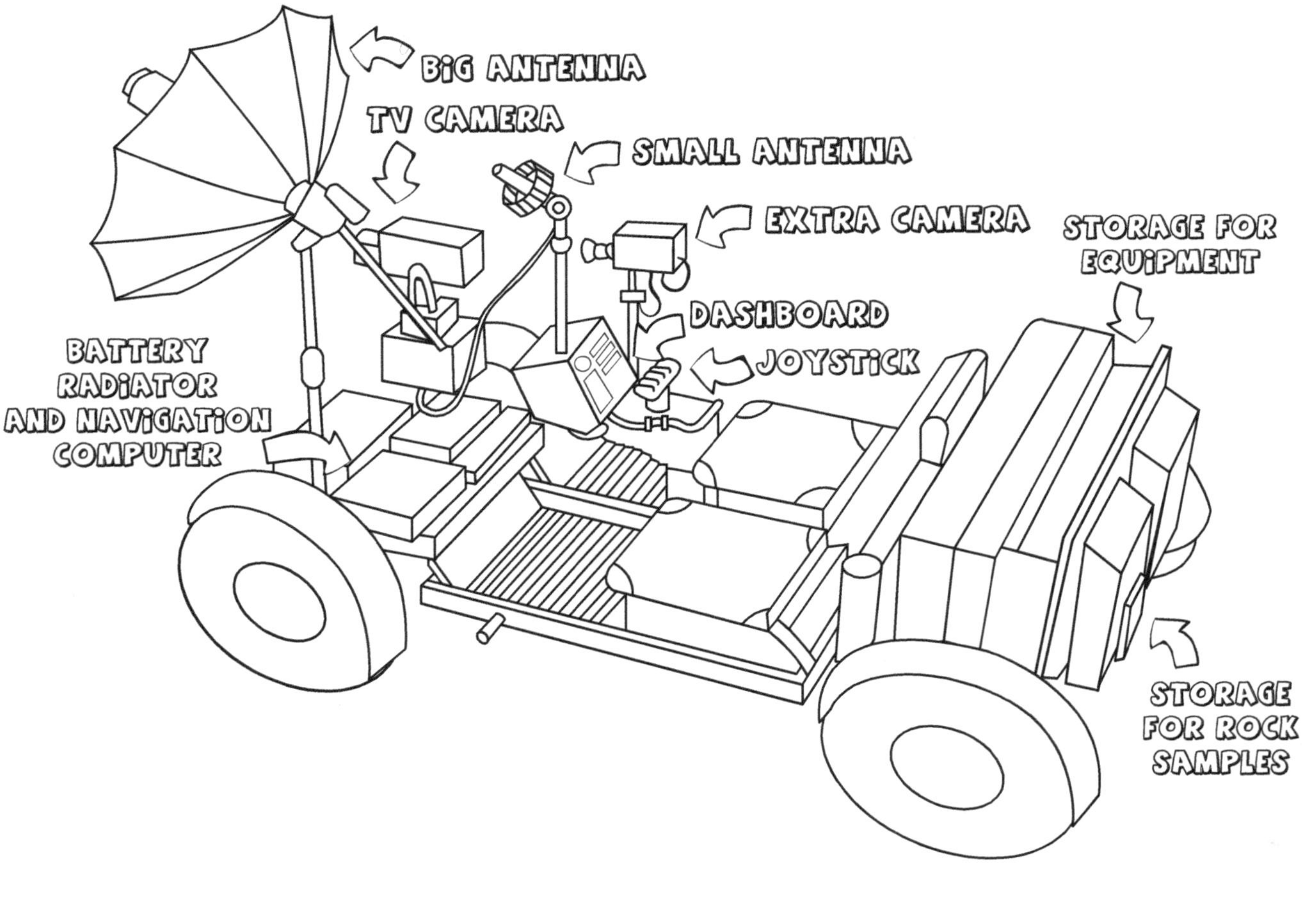

STEP 11: SOME MISSIONS TO THE MOON INCLUDED A MOON BUGGY THAT ASTRONAUTS COULD USE TO DRIVE AROUND AND EXPLORE MORE OF THE MOON.

GOING TO THE MOON
STEP 12

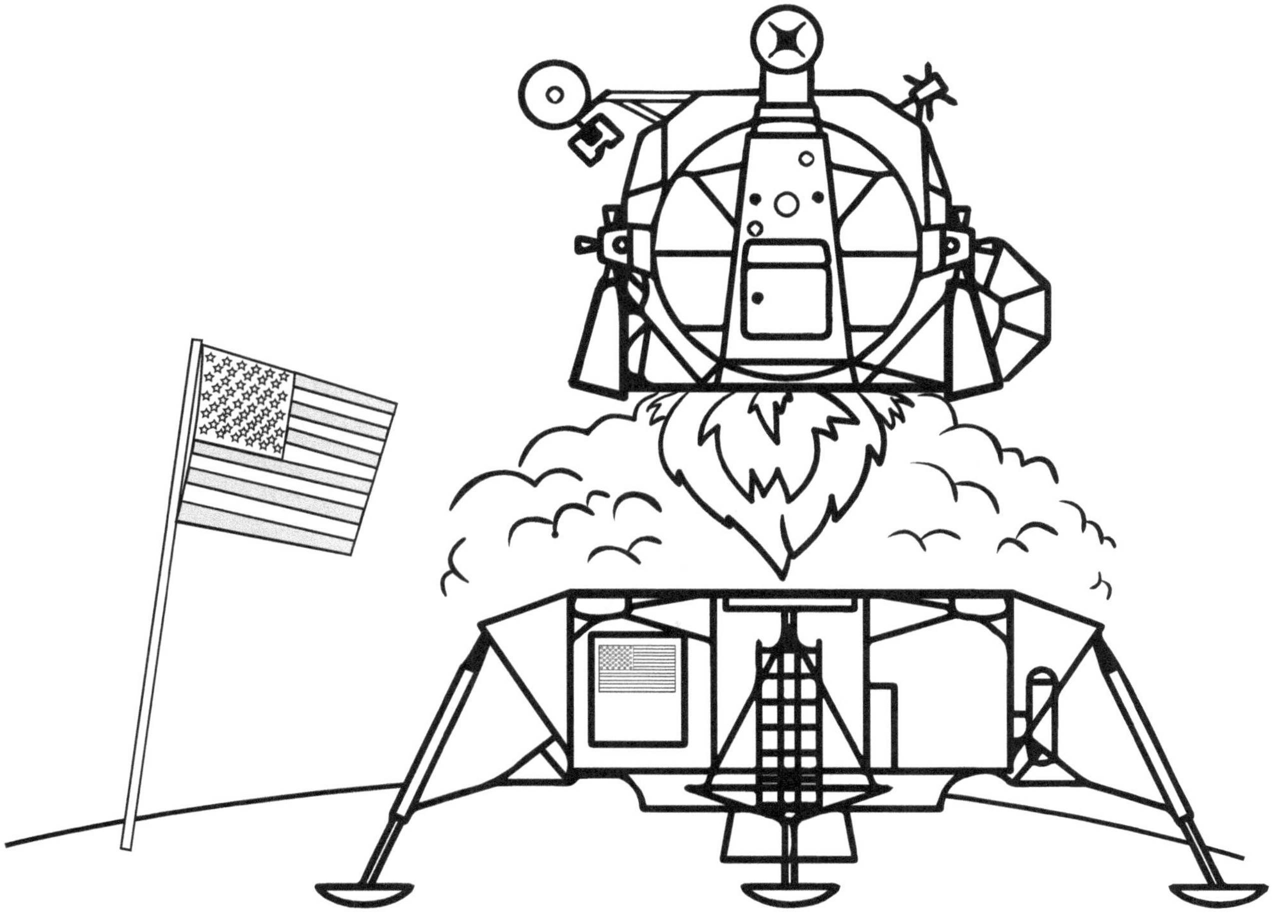

STEP 12: AFTER ALMOST A FULL DAY ON THE MOON, ASTRONAUTS CLIMB INTO THE TOP OF THE LUNAR LANDER (CALLED THE LUNAR ASCENT MODULE) AND LAUNCH OFF THE MOON.

GOING TO THE MOON
STEP 13

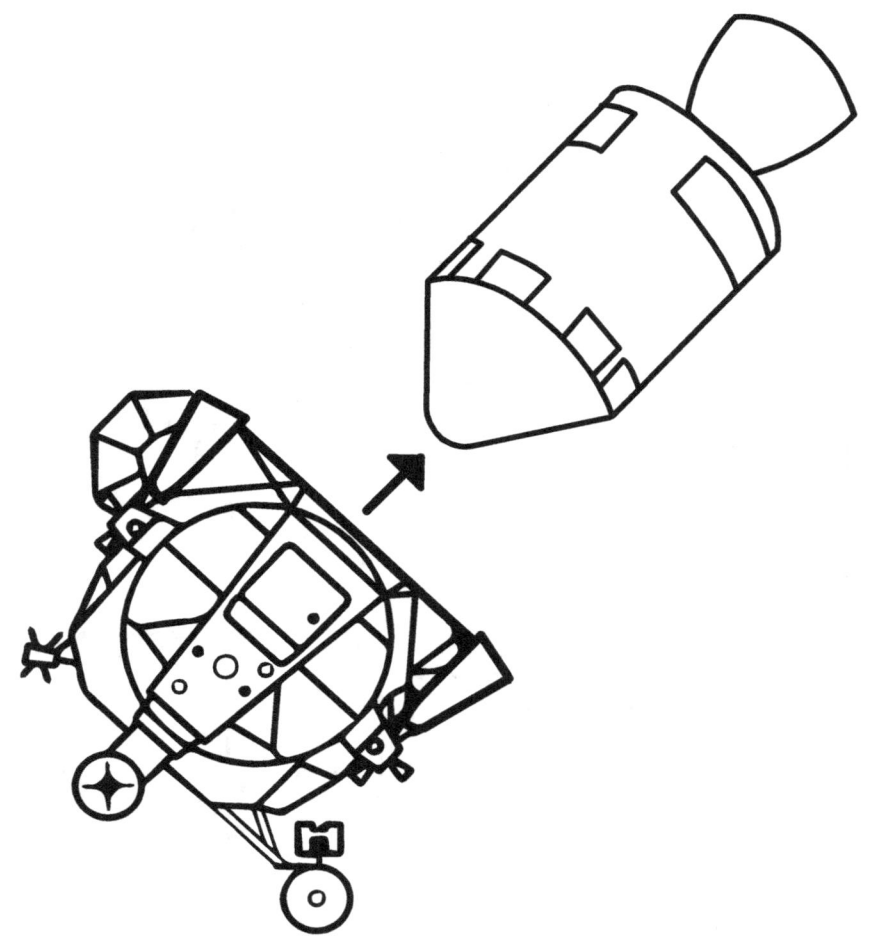

STEP 13: THE LUNAR ASCENT MODULE ACCELERATES INTO ORBIT AROUND THE MOON AND CONNECTS BACK TO THE COMMAND SERVICE MODULE.

GOING TO THE MOON
STEP 14

STEP 14: ALL THE ASTRONAUTS THEN MOVE INTO THE COMMAND SERVICE MODULE AND FIRE THE ROCKET ENGINE TO LEAVE MOON'S ORBIT AND GO BACK TOWARDS EARTH.

GOING TO THE MOON
STEP 15

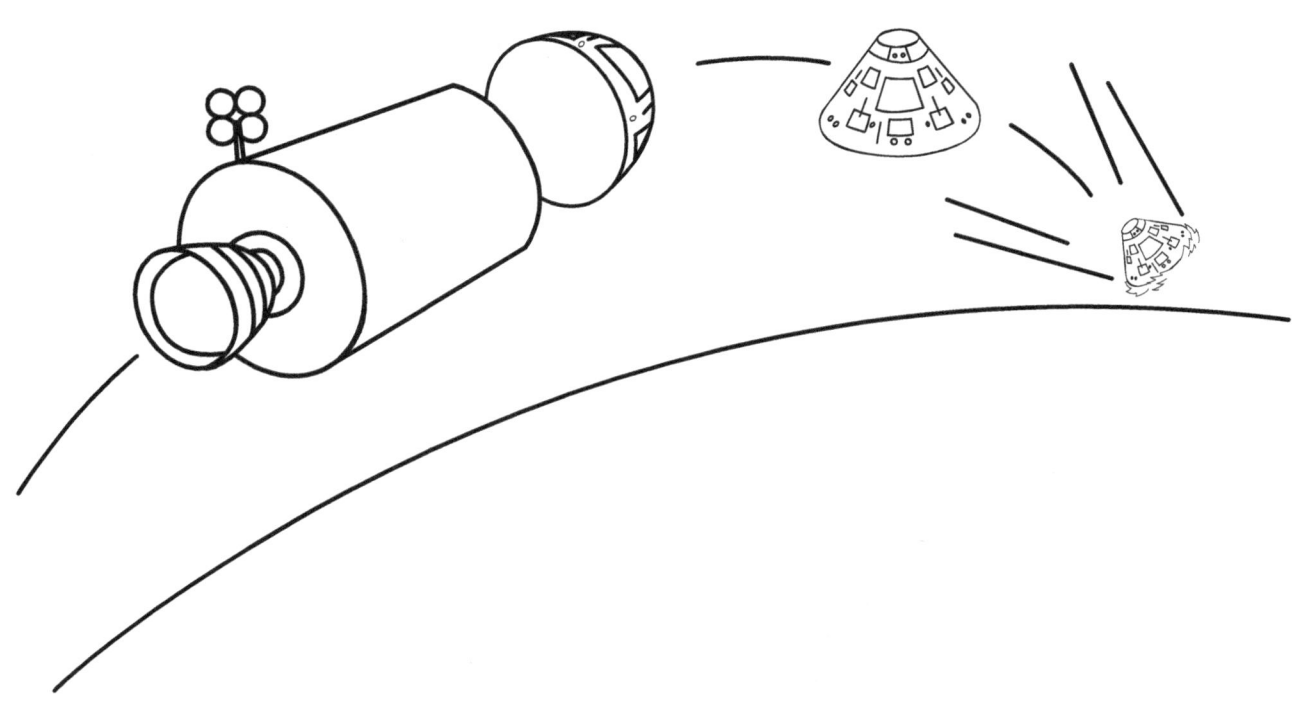

STEP 15: ONCE NEAR EARTH, THE COMMAND MODULE SEPARATES. THE AIR AROUND EARTH HELPS SLOW IT DOWN AS IT FALLS TOWARDS EARTH.

GOING TO THE MOON
STEP 16

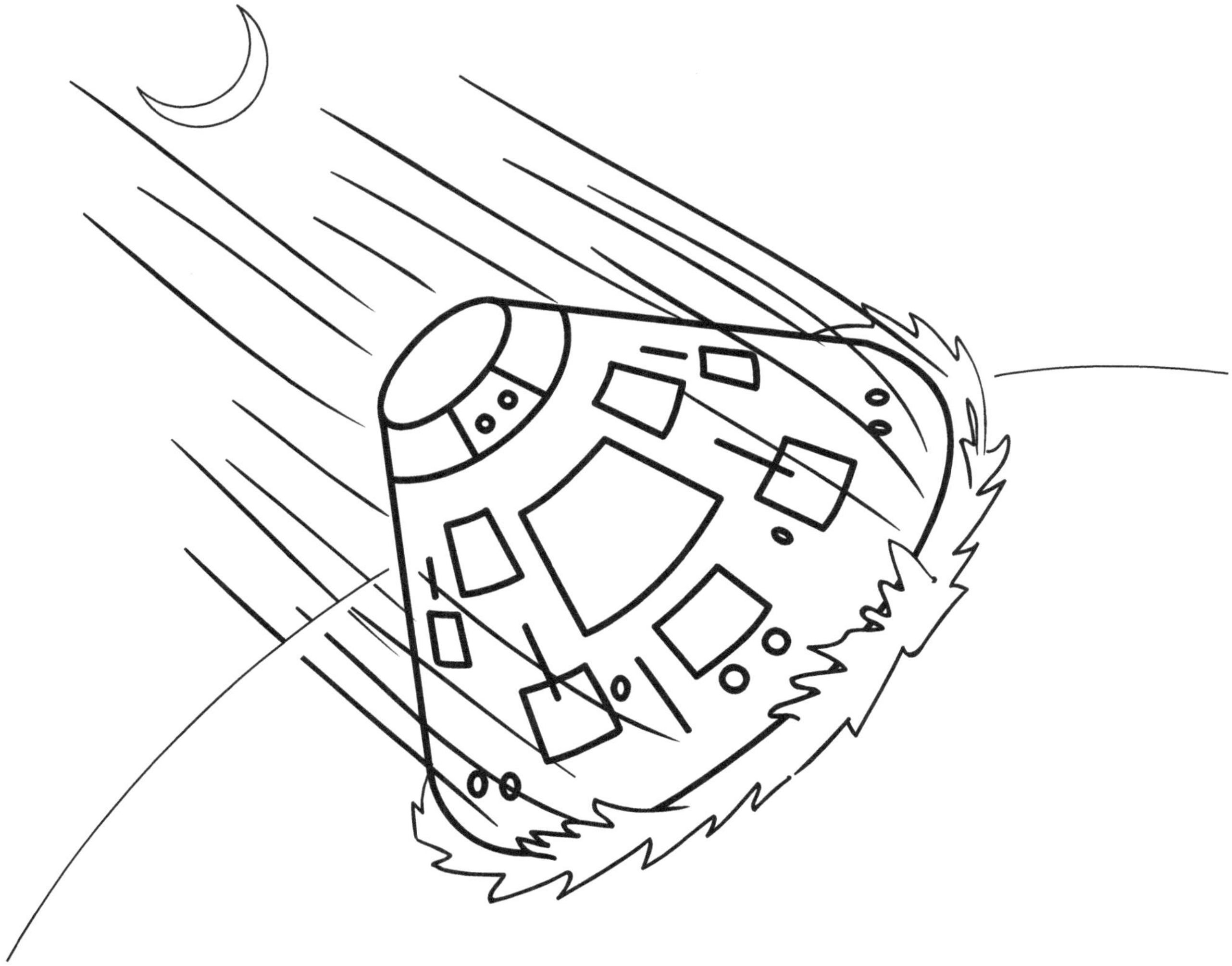

STEP 16: THE COMMAND MODULE IS MOVING SO FAST THROUGH THE AIR THAT IT GETS EXTREMELY HOT!

GOING TO THE MOON
STEP 17

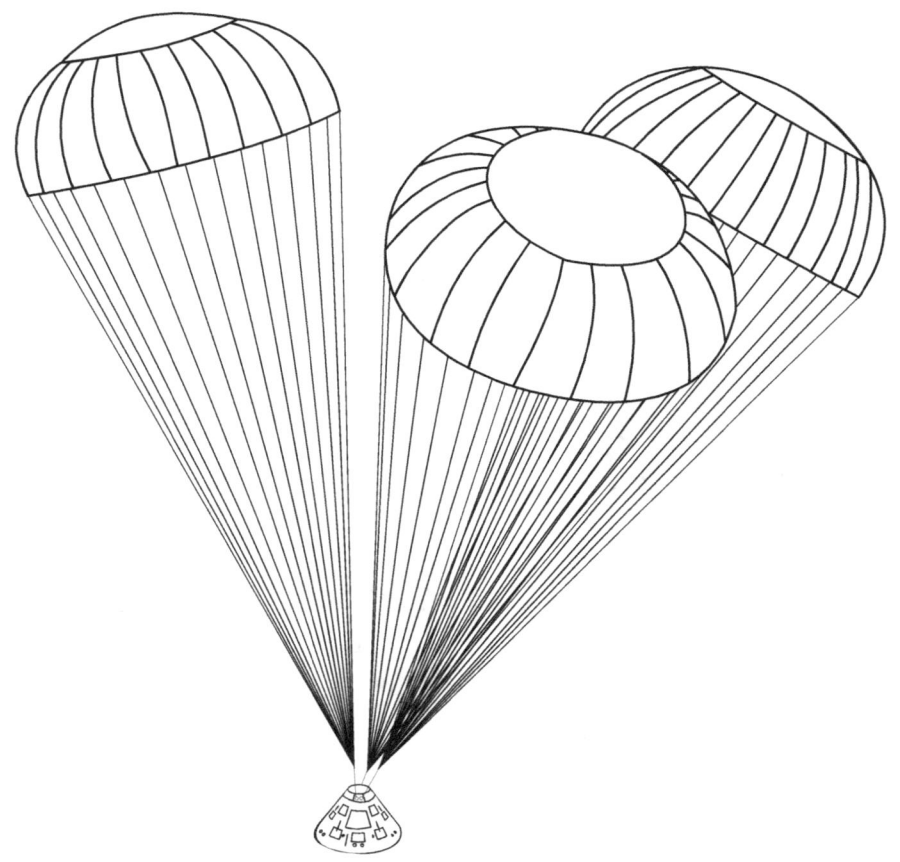

STEP 17: PARACHUTES THEN OPEN UP, AND THE MODULE SLOWLY FALLS TOWARDS EARTH.

GOING TO THE MOON
STEP 18

STEP 18: SPLASH DOWN! THE MODULE LANDS IN THE OCEAN AND FLOATS AS PEOPLE COME TO GET THE ASTRONAUTS OUT OF THE MODULE.

The Little Engineer Coloring Book: Space & Rockets

GOING TO THE MOON
STEP 19

STEP 19: SCIENTISTS THEN STUDY THE MOON ROCKS TO LEARN ABOUT HOW THE MOON WAS FORMED AND THE MATERIAL WITH WHICH IT IS MADE.

SPACE SHUTTLE

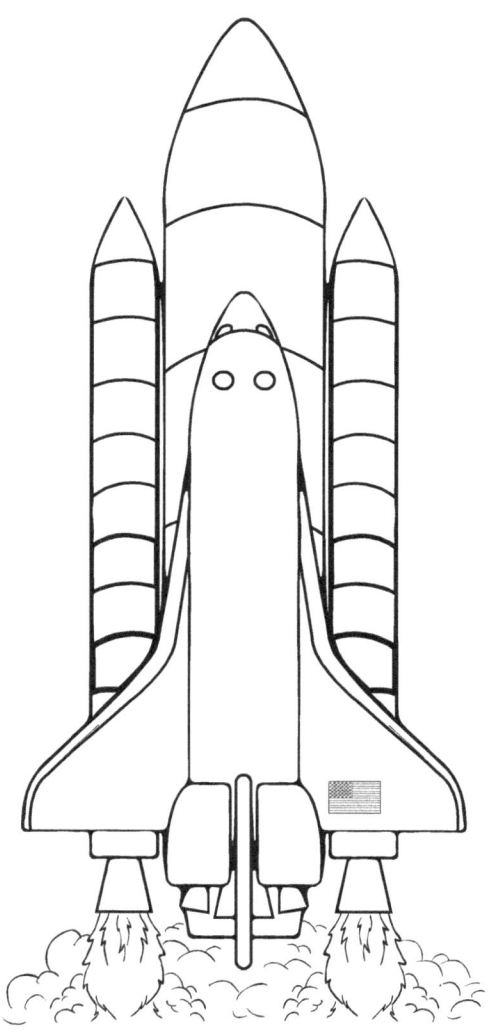

ONCE THE MOON MISSIONS WERE COMPLETED. THE SPACE SHUTTLE WAS CREATED SO IT COULD DELIVER THINGS TO SPACE, LAND LIKE A PLANE ON A RUNWAY AND THEN LAUNCH AGAIN.
THE SPACE SHUTTLE LAUNCHED WITH A HUGE FUEL TANK AND SIDE BOOSTERS.

SPACE SHUTTLE SIDE BOOSTERS

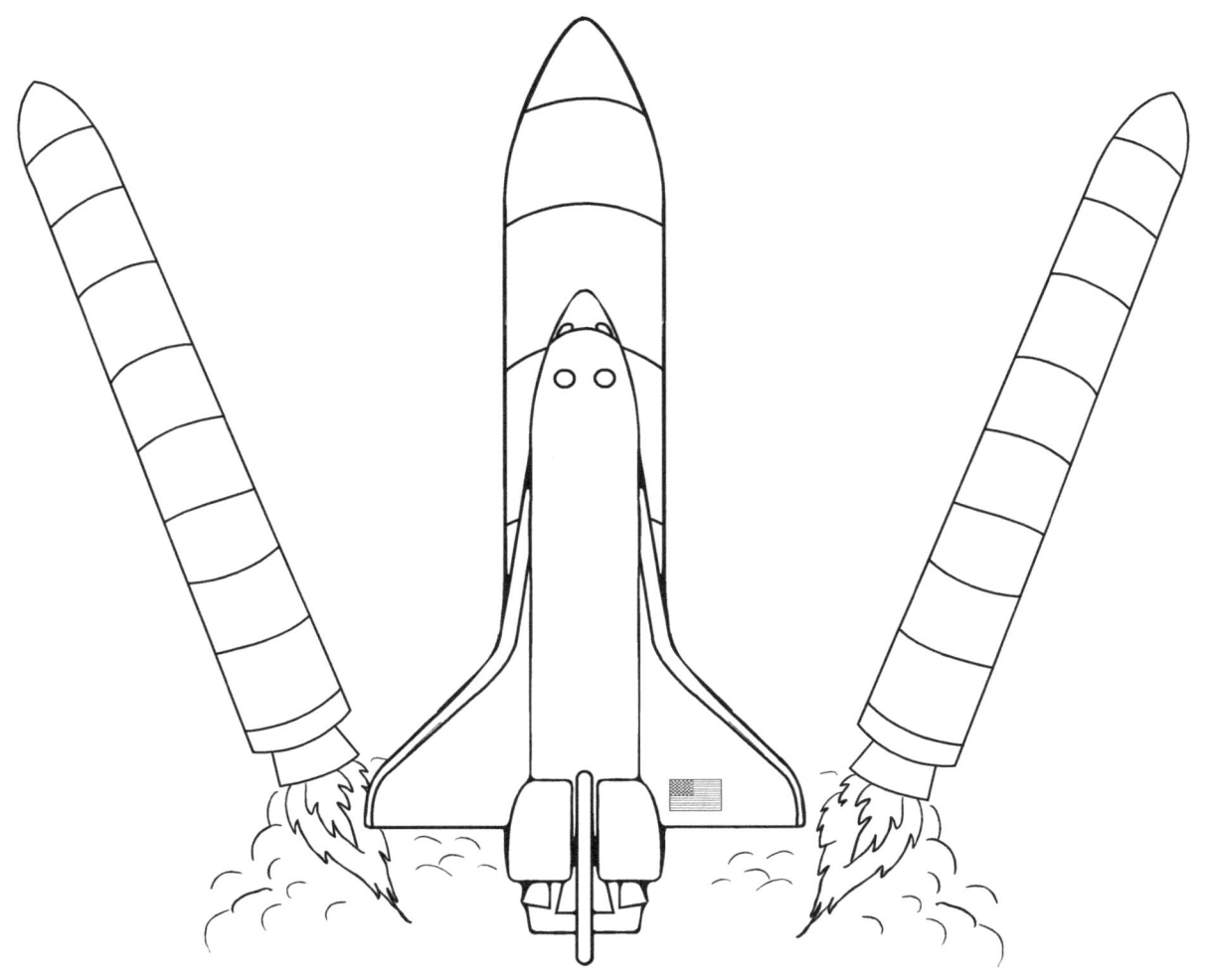

SIMILAR TO THE 1ST STAGE OF A ROCKET, THE SIDE BOOSTERS DISCONNECT FROM THE SHUTTLE DURING LAUNCH ONCE THEY ARE OUT OF FUEL.

SPACE SHUTTLE CARGO BAY

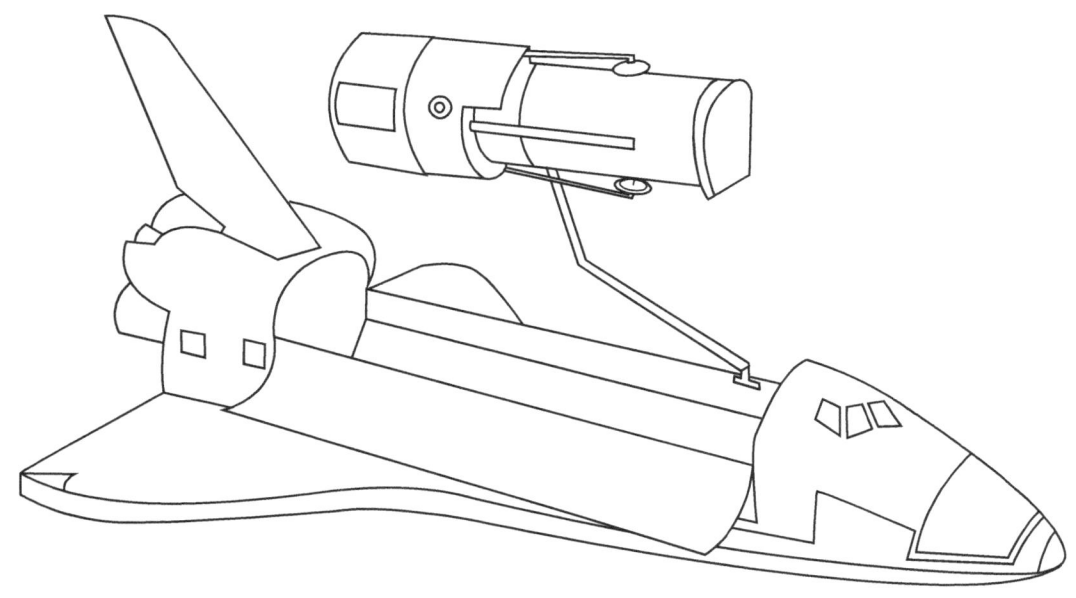

THE SPACE SHUTTLE HAS A HUGE CARGO BAY THAT ALLOWS IT TO CARRY ALL SORTS OF ITEMS INTO EARTH'S ORBIT. THE SPACE SHUTTLE LAUNCHED MANY SATELLITES AND PIECES OF THE INTERNATIONAL SPACE STATION.

SPACE SHUTTLE RETURN HOME

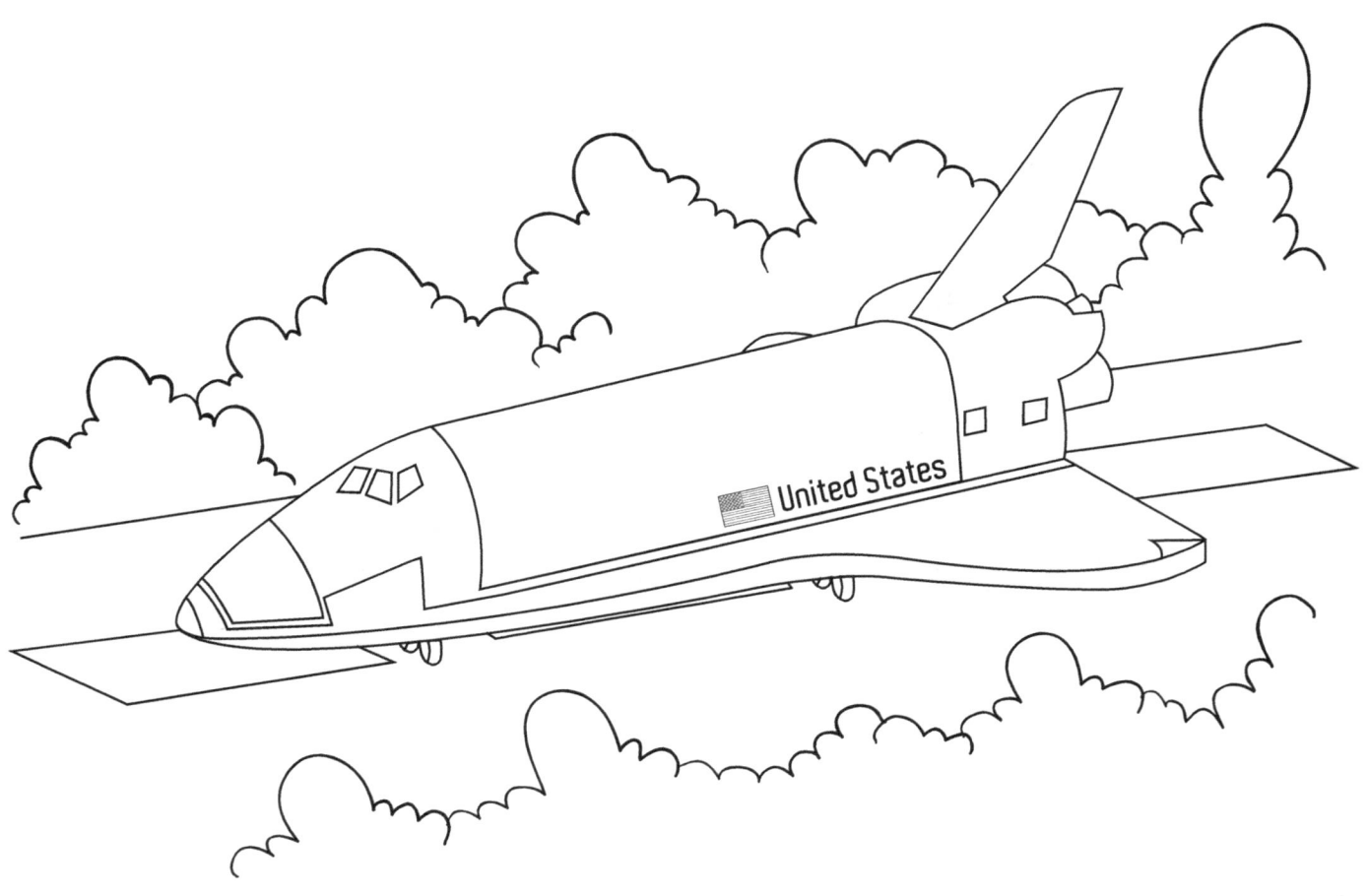

AFTER DELIVERING CARGO TO SPACE, THE SPACE SHUTTLE RETURNS TO EARTH AND LANDS JUST LIKE AN AIRPLANE.

SPACEX RE-USABLE ROCKETS

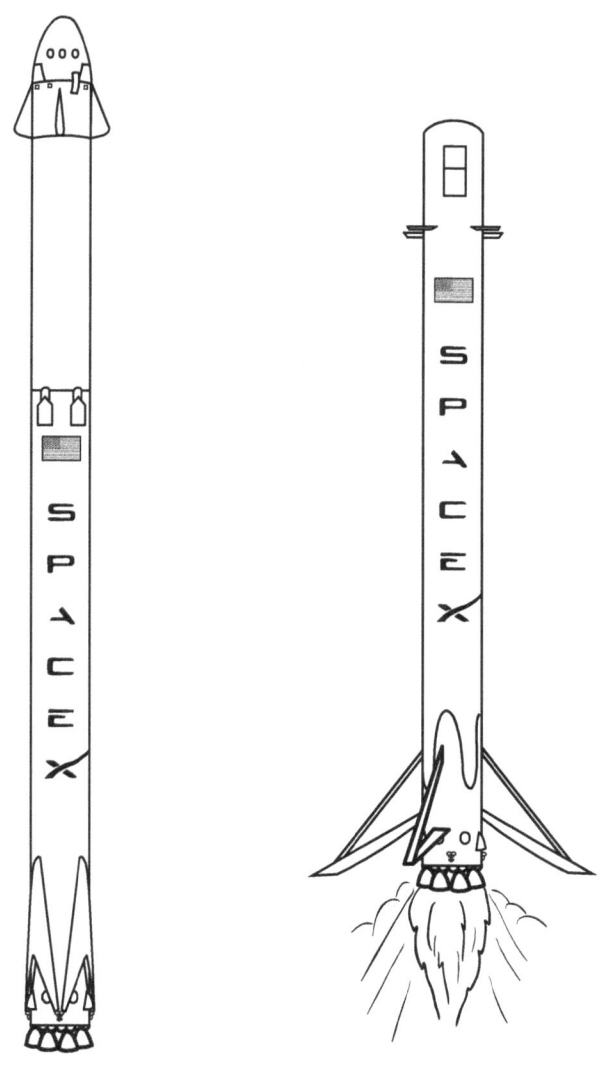

RECENTLY, A COMPANY CALLED SPACEX STARTED USING ROCKETS MORE THAN ONCE. AFTER LAUNCHING THE PAYLOAD, THE ROCKET WILL GO BACK TO EARTH AND LAND!

SPACEX
HOW THE ROCKET LANDS

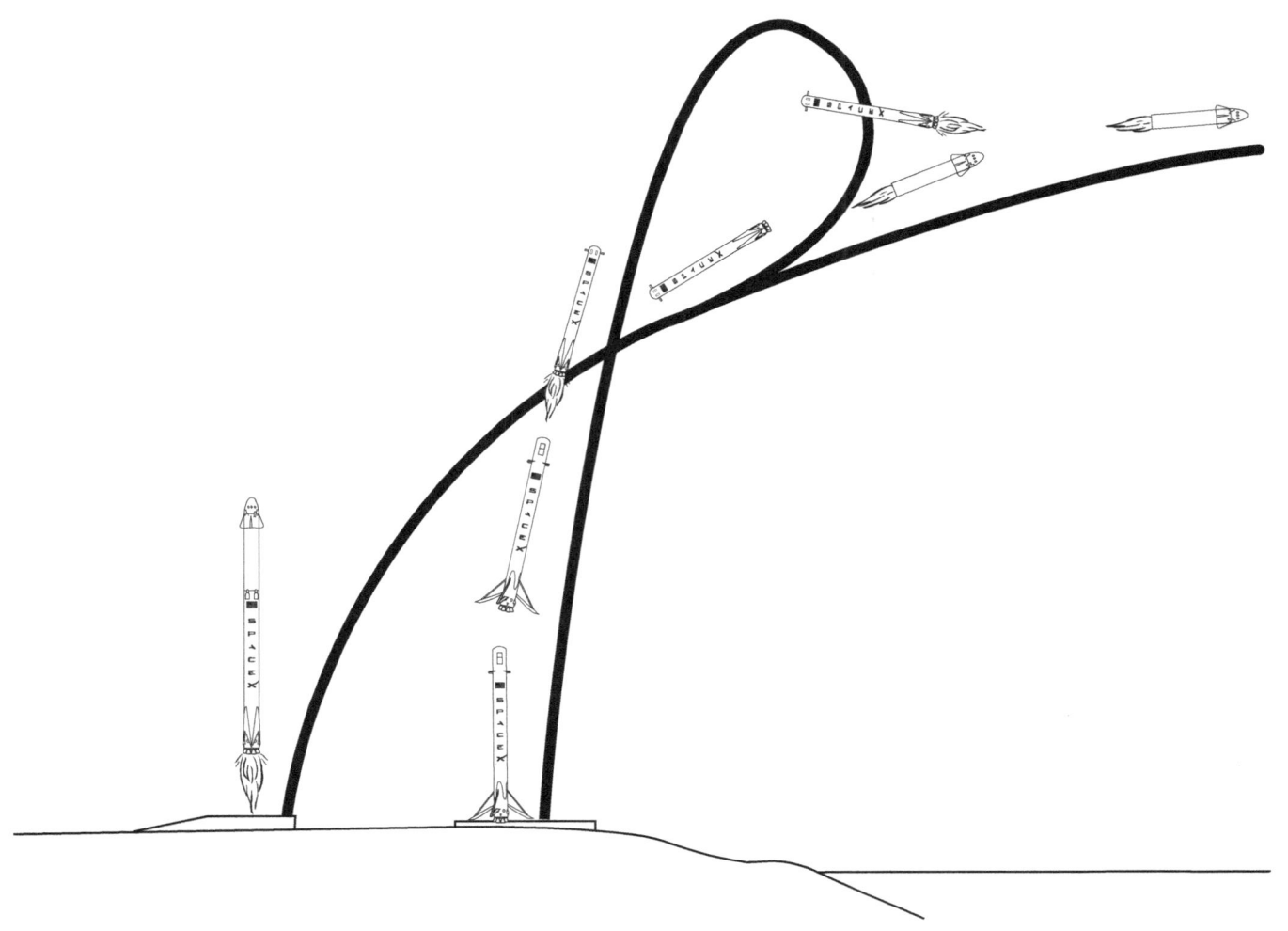

ONCE THE 1ST STAGE OF THE ROCKET SEPARATES FROM THE 2ND STAGE, THE 2ND STAGE WILL CONTINUE TOWARDS ORBIT, BUT THE 1ST STAGE THEN TURNS AROUND AND FLIES BACK TOWARDS A LANDING SPOT ON EARTH.

MODERN ROCKETS

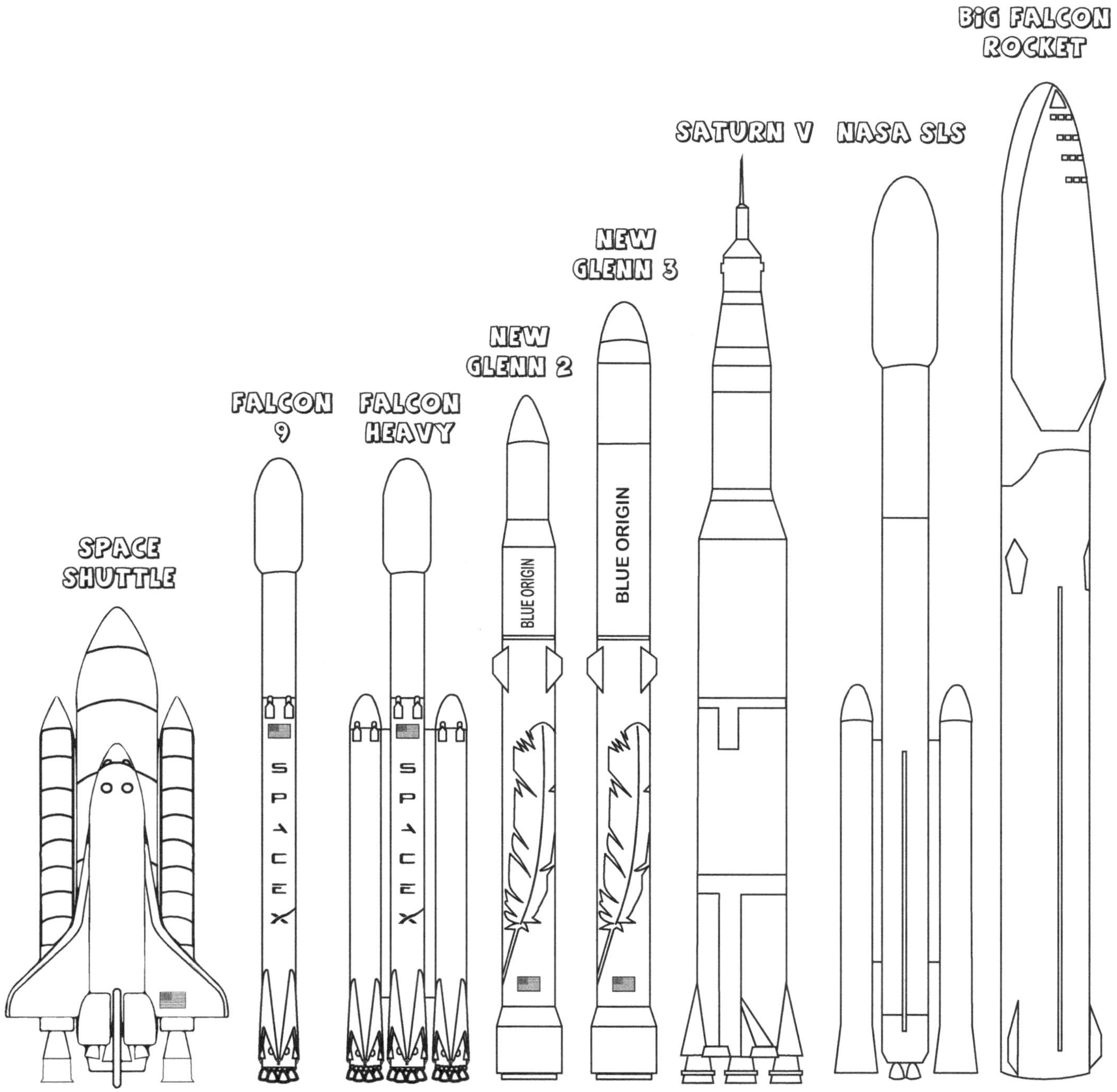

THERE ARE CURRENTLY MANY NEW ROCKETS JUST BECOMING AVAILABLE. THE SATURN V AND SPACE SHUTTLE ARE INCLUDED JUST FOR COMPARISON.

The Little Engineer Coloring Book: Space & Rockets

NEXT BIG ADVENTURE

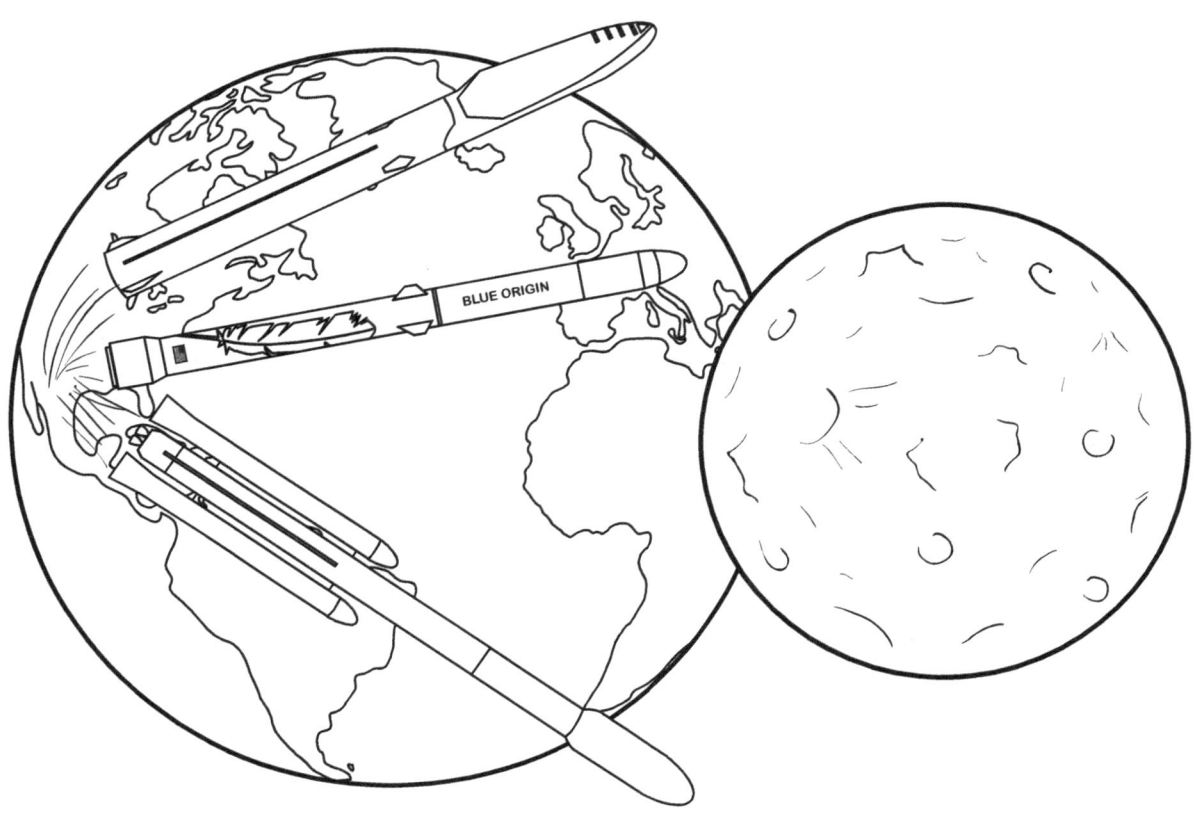

WHAT WILL OUR NEXT BIG ADVENTURE IN SPACE BE? WITH SO MANY GREAT NEW ROCKETS BECOMING AVAILABLE, SCIENTISTS AND ENGINEERS HAVE MANY GREAT OPTIONS.

REVISIT THE MOON

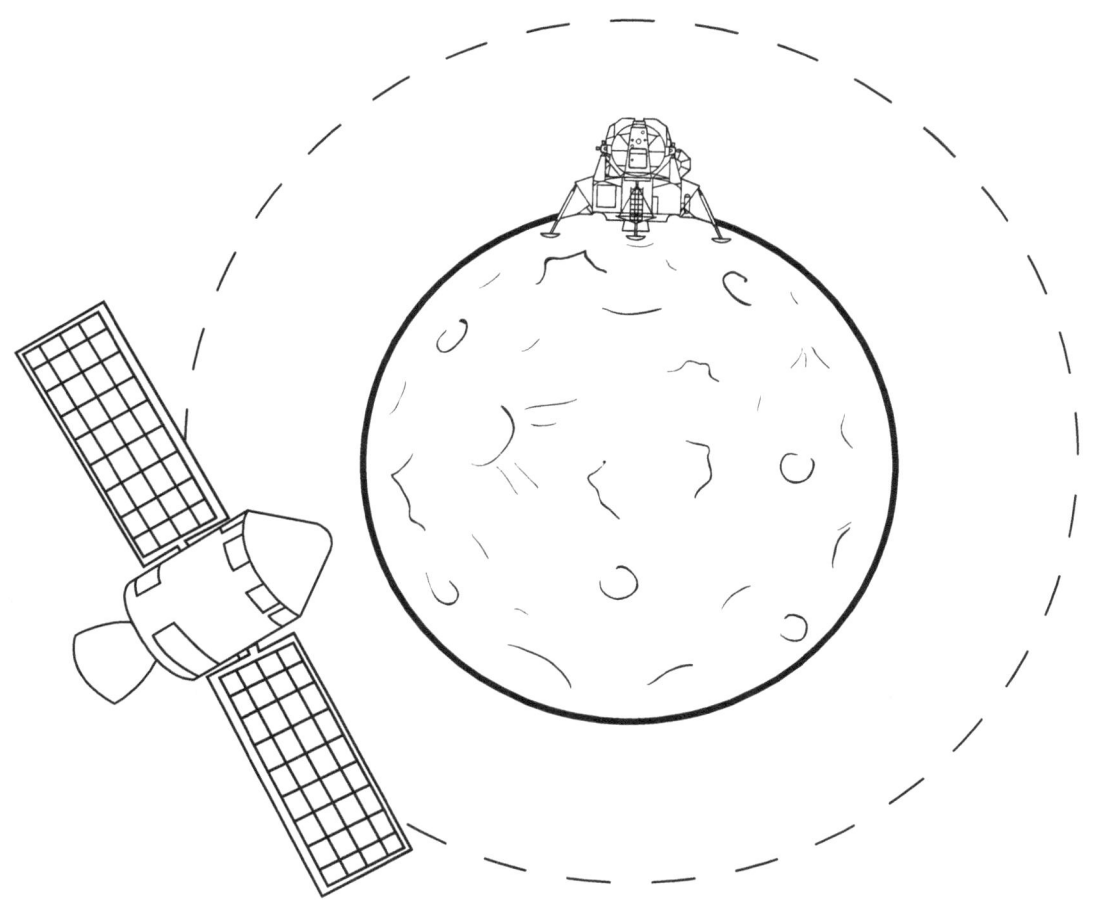

SCIENTISTS ARE ALREADY PUTTING A PLAN TOGETHER TO GO BACK TO THE MOON AND STAY THERE FOR LONG PERIODS OF TIME.

JOURNEY TO MARS

THE COMPANY, SPACEX, IS BUILDING THE LARGE BFR ROCKET WITH THE GOAL OF LANDING HUMANS ON MARS.

LAUNCH MORE SPACE PROBES

WITH MORE POWERFUL ROCKETS AVAILABLE, LARGER PROBES CAN BE SENT OUT INTO SPACE TO EXPLORE DISTANT PLANETS.

WHY EXPLORE SPACE?

SPACE IS A REALLY EXCITING ADVENTURE, BUT THERE ARE ALSO MANY USEFUL REASONS TO GO TO SPACE.

1. EARTH HAS A LIMITED AMOUNT OF RESOURCES, BUT SPACE IS UNIMAGINABLY BIG AND FULL OF RESOURCES. ONE NEW POSSIBILITY WILL BE TO CAPTURE AND USE THESE MATERIALS THAT WE MAY LACK HERE ON EARTH.

2. MANUFACTURING IN SPACE WOULD BE GREAT FOR THE EARTH. THIS WOULD ALLOW US TO MOVE HARMFUL FACTORIES FROM THE EARTH TO SPACE.

3. ROBOTIC MISSIONS TO MARS HAVE SHOWN IT ONCE WAS VERY SIMILAR TO EARTH. IT WILL BE HELPFUL TO STUDY MARS AND LEARN WHY IT IS NOW LIKE A DESERT PLANET. THIS WILL HELP US LEARN HOW TO KEEP EARTH BEAUTIFUL AND GREEN FOR A LONG TIME.

FUN IDEA FOR THE FUTURE!

IMAGINE IF PHONES WERE MADE IN SPACE FROM MATERIALS FOUND IN SPACE. THEN THE FINISHED PHONES WOULD BE SHIPPED DOWN TO EARTH.

THIS WOULD BE REALLY COOL, BUT ALSO GREAT FOR EARTH BECAUSE NONE OF THE TRASH OR POLLUTION FROM MAKING THE PHONE WOULD BE ON EARTH.

Congratulations! Training Complete

You are now certified and have successfully completed your Space and Rockets Training!

Go to: thelittleengineerbooks.com/tle-rockets-certificate or scan the QR code, and enter your email to get a ready-to-print certificate

SPECIAL PREVIEW

SETH MCKAY

Check out this short preview of another fun coloring book!

ENGINE ACCESSORIES

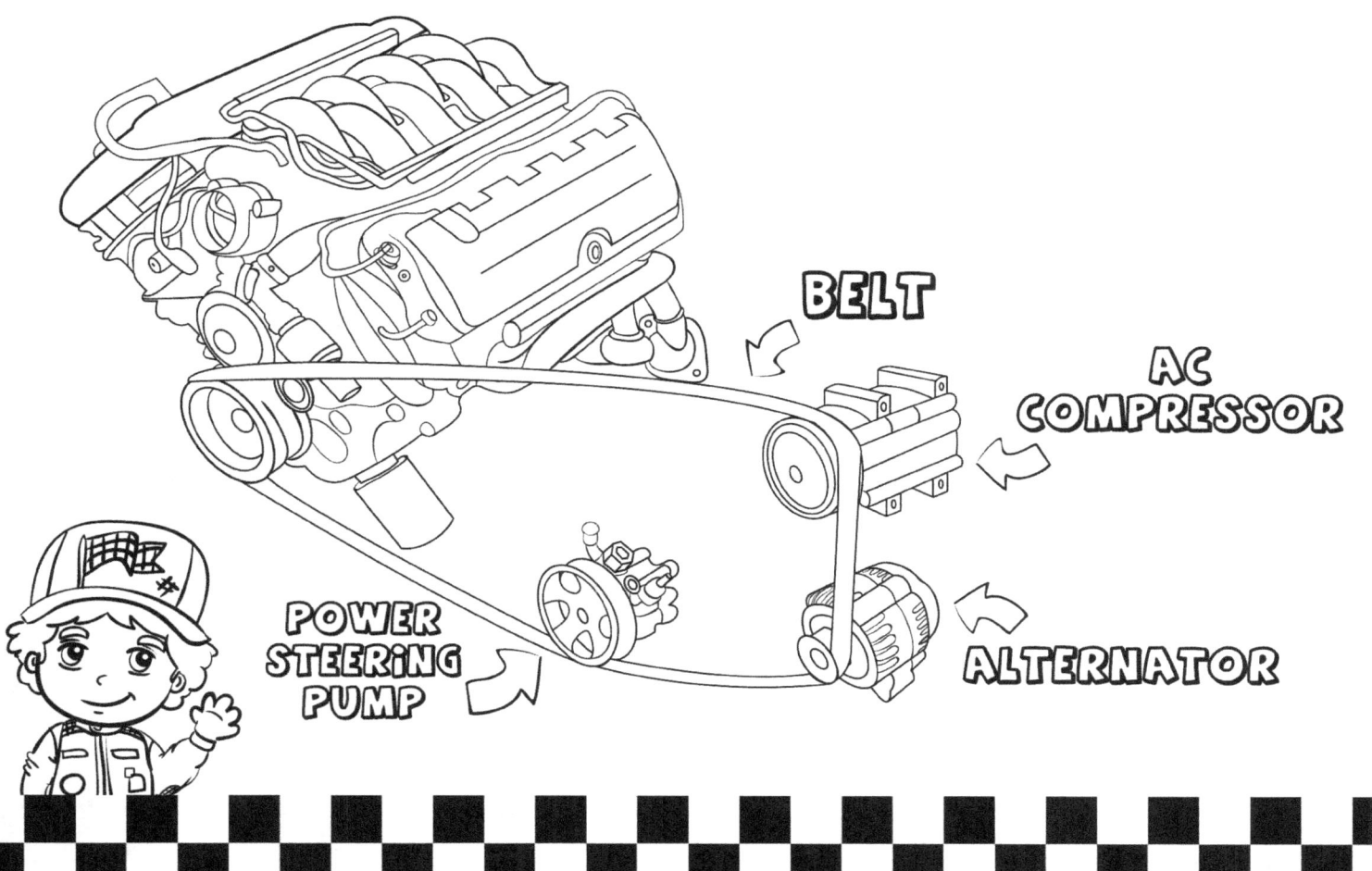

THE ENGINE ACCESSORIES ARE POWERED BY A BELT ON THE ENGINE. THE BELT SPINS THE WHEELS ON THE ACCESSORIES WHEN THE ENGINE IS ON.

- THE ALTERNATOR IS THE BATTERY CHARGER FOR THE CAR. IT IS A SMALL POWER GENERATOR THAT MAKES SURE YOUR CAR HAS PLENTY OF ELECTRICITY FOR LIGHTS, SPARK PLUGS, THE RADIO AND MORE.
- THE AC COMPRESSOR HELPS THE AC SYSTEM WORK SO THE AIR IS NICE AND COLD INSIDE THE CAR.
- THE POWER STEERING PUMP MAKES IT EASIER TO TURN THE STEERING WHEEL.

RADIATOR

ENGINES GET VERY HOT.

- A RADIATOR HELPS KEEP THE ENGINE FROM GETTING TOO HOT.
- A WATER PUMP MOVES A SPECIAL LIQUID CALLED COOLANT THROUGH THE ENGINE AND THEN THROUGH A RADIATOR.

2 DIFFERENTIALS

ON TRUCKS, YOU CAN SOMETIMES SEE 2 DIFFERENTIALS. THIS MEANS THE TRUCK HAS 4-WHEEL DRIVE. SOME CARS HAVE 4-WHEEL DRIVE, BUT IT IS HARD TO SEE THE DIFFERENTIAL BECAUSE THE CAR IS CLOSE TO THE GROUND.

TWIN TURBOCHARGERS

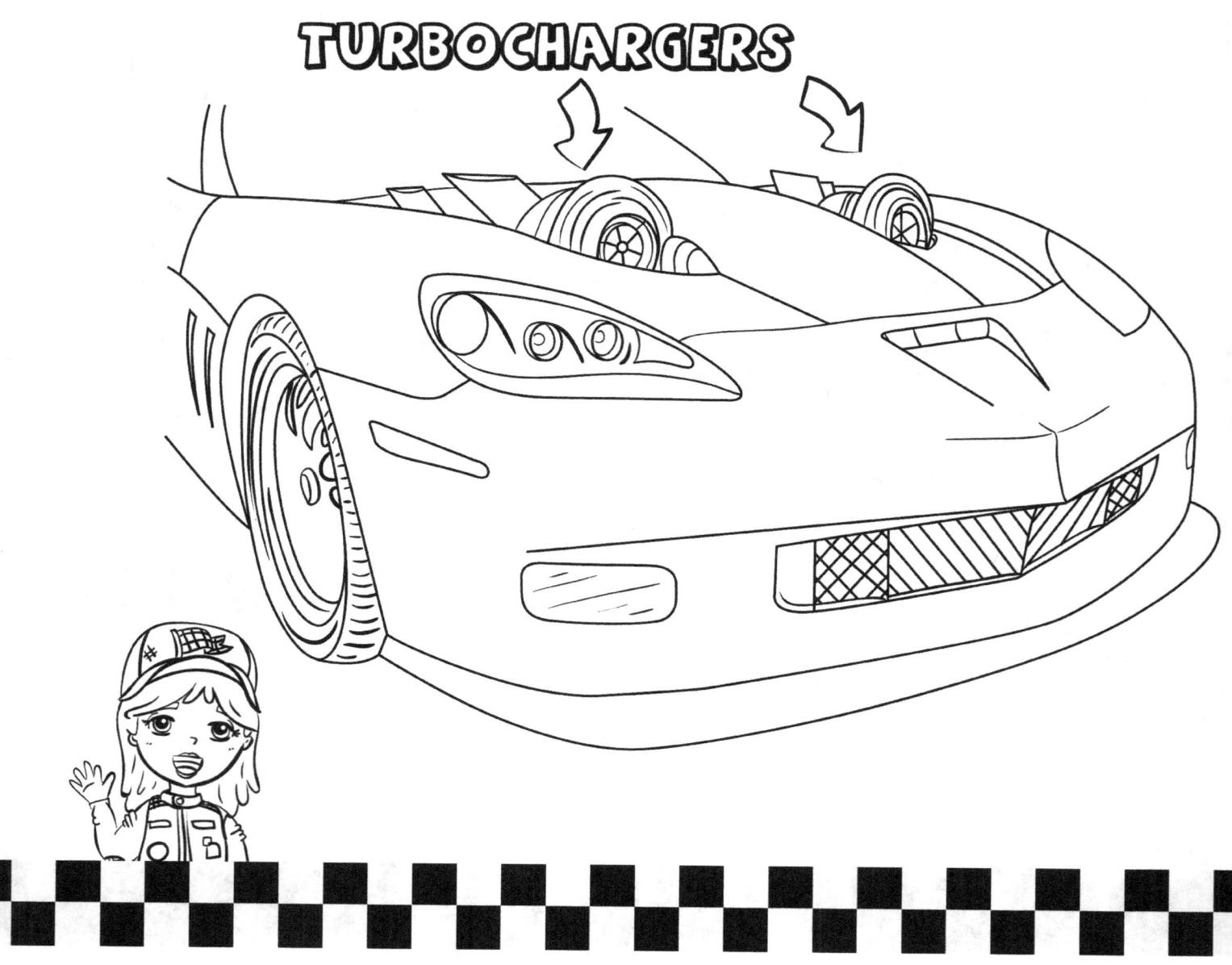

THIS CAR HAS 2 TURBOCHARGERS!

MOST TURBOCHARGERS ARE UNDER THE HOOD AND HARD TO SEE, BUT SOME CARS HAVE THEM STICKING OUT OF THE HOOD WHICH IS REALLY COOL!

We hope you enjoyed the book!
Contact us anytime at CreativeIdeasPublishing.com

We are a US based publisher that consist of parents and teachers. We try our best to make products that our kids will love and we hope your kids love them too!

Ask your bookstore for more great titles from Creative Ideas Publishing!

   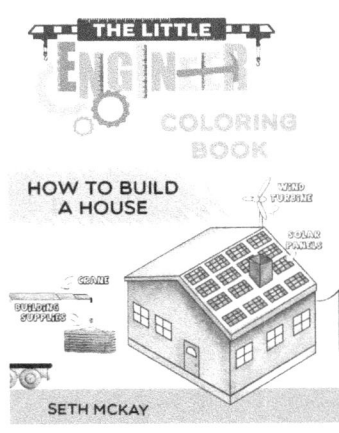

www.ingramcontent.com/pod-product-compliance
Lightning Source LLC
Chambersburg PA
CBHW081756100526
44592CB00015B/2458